An Introduction to Marine Science

TERTIARY LEVEL BIOLOGY

A series covering selected areas of biology at advanced undergraduate level. While designed specifically for course options at this level within Universities and Polytechnics, the series will be of great value to specialists and research workers in other fields who require a knowledge of the essentials of a subject.

Recent titles in the series:

Biology of Reptiles	Spellerberg
Biology of Fishes	Bone and Marshall
Mammal Ecology	Delany
Virology of Flowering Plants	Stevens
Evolutionary Principles	Calow
Saltmarsh Ecology	Long and Mason
Tropical Rain Forest Ecology	Mabberley
Avian Ecology	Perrins and Birkhead
The Lichen-Forming Fungi	Hawksworth and Hill
Plant Molecular Biology	Grierson and Covey
Social Behaviour in Mammals	Poole
Physiological Strategies in Avian Biology	Phillips, Butler and Sharp
An Introduction to Coastal Ecology	Boaden and Seed
Microbial Energetics	Dawes
Molecule, Nerve and Embryo	Ribchester
Nitrogen Fixation in Plants	Dixon and Wheeler
Genetics of Microbes (2nd edn.)	Bainbridge
Seabird Ecology	Furness and Monaghan
The Biochemistry of Energy Utilization in Plants	Dennis
The Behavioural Ecology of Ants	Sudd and Franks
Anaerobic Bacteria	Holland, Knapp and Shoesmith

TERTIARY LEVEL BIOLOGY

An Introduction to Marine Science

Second Edition

P.S. MEADOWS, MA, BA, FZS
J.I. CAMPBELL, BSc, FRES
Department of Zoology
University of Glasgow

Blackie

Glasgow and London

Halsted Press, a Division of

John Wiley and Sons

New York

Blackie and Son Ltd,
Bishopbriggs, Glasgow G64 2NZ
7 Leicester Place, London WC2H 7BP

Published in the USA by
Halsted Press, a Division of
John Wiley and Sons Inc., New York .

British Library Cataloguing in Publication Data

Meadows, P S
    An introduction to marine science.————
    2nd ed.———— (Tertiary level biology).
    1. Oceanography
    I. Title   II. Campbell, J I   III. Series
    551.46   GC11.2

    ISBN 0–216–92267–4
    ISBN 0–216–92268–2 Pbk

Library of Congress Cataloging-in-Publication Data

Meadows, P S
    An introduction to marine science / P S Meadows, J I Campbell. —
—2nd ed.
        p.   cm. — (Tertiary level biology)
    Bibliography: p.
    Includes index.
    ISBN 0-470-20951-8.   ISBN 0-470-20952-6 (pbk.)
    1. Marine biology.   2. Marine sciences.   I. Campbell, J.
I. (Janette I.)   II. Title.   III. Series.
QH90.M43 1988
574.92—dc19   87-20603
                                                       CIP

Phototypeset at Thomson Press (India) Ltd., New Delhi.
Printed in Great Britain by Bell & Bain Ltd.

Preface

It is now nine years since the first edition appeared and much has changed in marine science during that time. For example, satellites are now routinely used in remote sensing of the ocean surface and hydrothermal vents at sea floor spreading centres have been extensively researched.

The second edition has been considerably expanded and reorganised, and many new figures and tables have been included. Every chapter has been carefully updated and many have been rewritten. A new chapter on man's use of the oceans has been included to cover satellites and position fixing, renewable energy sources in the sea, seabed minerals, oil and gas, pollution and maritime law. In this edition we have also referred to a number of original references and review articles so that readers can find their way into the literature more easily. As in the first edition, PSM has been mainly responsible for the text and JIC for the illustrations, although each has responded to advice from the other and also from many colleagues. In this context readers should note that the illustrations form an integral and major part of the book. The text will almost certainly be too concise for many readers if they do not study the illustrations carefully at the same time.

The book has been written as an introductory text for students, although it can serve anyone who is beginning a study of the sea. The examples with which we have illustrated the text are drawn from a wide range of seas. For instance, we have considered coral reefs and mangrove swamps in the Indo-Pacific and Atlantic, the Great Barrier Reef on the north-east coast of Australia, worldwide fisheries, fish farming in the Far East, and the mining of manganese nodules in the Pacific. As in the first edition, we have attempted to give the book a more representative coverage of the study of the sea at an introductory level than is currently available.

We gratefully acknowledge the following past and present members of PSM's research group for much stimulating discussion, many of whom are now in senior posts in Britain or abroad: John Anderson, Andrew Boney, Robin Bruce, Mark Corps, Elizabeth Deans, Farage Eddeb, Cahit Erdem,

Andrew Girling, Mohammad Hariri, Eman Hilal, Samira Hussain, Michelle Kirkham, Hadi Mgherbi, Robert Millar, Kenneth Mitchell, Brian Mullins, John Murray, Salvador Ramirez, Alan Reichelt, Amer Ruagh, Ibrahim Saleh, Peter Shand, Joseph Tait, Azra Tufail, James Waterworth, and James Wilson. We are particularly grateful to Pat MacLaughlin (manuscript preparation), Alan Reichelt (meiofauna, P/B ratios) and Azra Tufail (phytoplankton, primary production). We would also like to thank the following colleagues for advice and help: Martin Angel, Jim Atkinson, Peter Barnett, Murdoch Baxter, Brian Bayne, Don Boney, Richard Crawford, David Cronan, Keith Dyer, Graham Durant, Peter Edmunds, George Farrow, John Gage, Andrew Gooday, Fred Grassle, Bob Hessler, Atig Huni, Holgar Jannasch, Colin Little, Alastair McIntyre, Paul Mayo, Geoff Moore, Trevor Norton, David Paterson, Gordon Petrie, Don Rhoads, Tony Rice, John Robinson, Howard Sanders, Rudolf Scheltema, Warren Smith, Peter Spencer Davies, John Spicer, Dick and Megumi Strathman, Alan Taylor, Don Tiffin, Mohammad Tufail, Paul Tyler, David Wethey, Dennis Willowes and Sally Woodin.

PSM
JIC

THIS SECOND EDITION IS DEDICATED TO OUR RESPECTIVE PARENTS

Contents

chains and webs. Phytoplankton growth, light and depth, compensation and critical depths. Seasonal plankton cycle, grazing and vertical
mixing. Geographical variations in seasonal cycles and productivity.
Zooplankton feeding. Plankton patchiness. Plankton indicator
species. Neuston. Pleuston. Epipelagic organisms, nekton. Mesopelagic, bathypelagic and abyssopelagic animals. Deep water sampling
methods. Whales, migration, krill. Sea birds: feeding, migration.

A minimum potential energy earth. Slow vertical movement of the
earth's surface, isostasy. Evidence from gravity anomalies, seismic
reflection and refraction, and continuous reflection profiling. Precipitation and evaporation, continent erosion and isostatic adjustment. Sea
floor spreading, plate tectonics. Palaeomagnetism, continental drift,
South America/Africa coastline fit. Mid-oceanic ridges, magnetic
reversal and sea floor spreading, *Glomar Challenger*, Deep Sea
Drilling Project. Lines of earthquake and volcanic activity, ocean
trench and mountain formation. Formation and destruction of the
earth's crust. Continental shelves and dams, ice age sea level.

A. The sedimentary environment. Continental shelf sediments, bottom
currents. Bottom topography, abyssal plains and hills. Sea mounts and
guyots. Microtopography. Turbidity currents and sediment transport.
Deep sea terrigenous and pelagic sediments. Manganese nodules.
Sedimentation rates in the deep sea. Sediment characteristics, shear
strength, erosion. Sediment diagenesis. B. The benthos. Infaunal and
epifaunal benthos, size classification. Macrofauna, food, bioturbation.
Meiofauna. Macro-algae, communities, production, production/
biomass ratio. Sea urchins, otters, and seaweeds. Continental shelf
benthic communities. Larvae of benthic invertebrates. Diversity and
stability in benthic communities. Ecological effects of increasing water
depth. Deep-sea benthic animals. Adaptations to deep sea life.
Hydrothermal vents, vent fauna.

Tidal ranges and levels. Supralittoral, intertidal and sublittoral zones.
Rocky, sandy and muddy shores. Wave action. Breaker, surf, swash
zones. Coastal currents. Longshore and rip currents. Longshore sand
transport. Changing sea level. Geomorphological beach zones. Sandy
and muddy shores. Particle size. Origin of sand on beaches. Summer
and winter beach profiles. Rocky shores. Exposed and sheltered sites.
Animal and plant zonation. Animal and plant zonation on sandy and
muddy shores. Causes of zonation on shores. Energy transfer, nutrient
cycling, food webs. Adaptations to intertidal conditions. Behavioural
responses and habitat selection.

Coastline modification. Sediment load. Patterns of estuarine circulation, bores. Sediment formation. Seasonal effects on water flow and
sedimentation. Nutrient input, output and cycling. Freshwater,
brackish-water and marine animals and their overlap. Birds. Food

CHAPTER ONE

THE SEA AS AN ENVIRONMENT

Distribution of land and sea on the earth's surface

The world's oceans cover about 70% of the earth's surface, and in places are deeper than 10 000 m. They are divided into the Pacific, the Atlantic, Indian, Arctic and Antarctic Oceans, as well as smaller areas like the North Sea between Britain and the continent of Europe, and the Mediterranean. These oceans, of course, are connected to each other, and can be looked upon as one super-ocean with many branches (Figure 1.1). The Pacific is the largest, and covers one-third of the earth. More of the earth's surface is covered by sea water in the southern hemisphere than in the northern, about 80% compared with about 60%; in fact, between latitude 55° and 65° south, that is, just south of Cape Horn, there is a complete ring of sea water

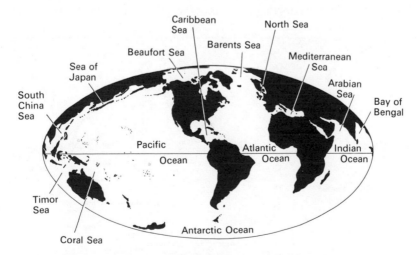

Figure 1.1 Major world oceans and some of the smaller seas. Some of the smaller seas are classified as part of the major oceans; for example, the Arabian Sea and the Bay of Bengal are part of the Indian Ocean, and the Coral Sea is part of the Pacific.

Table 1.1 Areas and volumes of the world's oceans

A. Areas of sea and land on earth

	World overall	Northern hemisphere	Southern hemisphere
% sea	70	60	80
% land	30	40	20

B. Depths of the oceans

	Overall average	Atlantic Ocean average	Indian Ocean average	Pacific Ocean average
Depth in metres	3800	3900	4000	4300

around the globe unbroken by land. Compared with mean sea level, the average depth of the oceans is about 3800 m, while the average height of the land is about 840 m, so the oceans are much deeper than the land is high (Table 1.1).

The main oceanic *biographic regions* are determined by temperature, and although not as obvious as the vegetation belts on land, are fairly well defined (Figure 1.2). For example the northern coasts of Britain are in the Atlantic Boreal region and the southern coasts are in the Atlantic Warm Temperate, while the northern coasts of Australia are in the Tropical Indo-

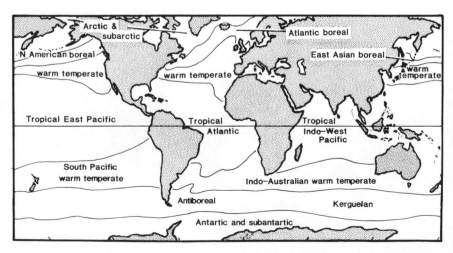

Figure 1.2 Major marine biogeographic areas of the world's oceans (after Couper, 1983).

West-Pacific and the southern coasts are in the Indo-Australian warm temperate region.

Depth and extent of different regions of the oceans

How does the depth of the ocean vary from place to place, and how are these variations related to the land? Firstly, most of the earth's surface is concentrated at two levels: these levels lie between 6000 and 4000 m below sea level, and between 500 m below sea level and 1000 m above sea level (Figure 1.3). The first includes the major ocean basins and the second a large part of most continents with the shallow seas surrounding them. Within these two major levels there is considerable complexity.

In most oceans we can distinguish three major levels or physiographic divisions: the continental margin, the ocean basin floor, and the mid-oceanic ridge (Figure 1.4).

(i) *The continental margins*

Around most continents there is a shallow shelf, the *continental shelf* (Figure 1.5). It slopes very gently at about 0.1° to the horizontal, from the

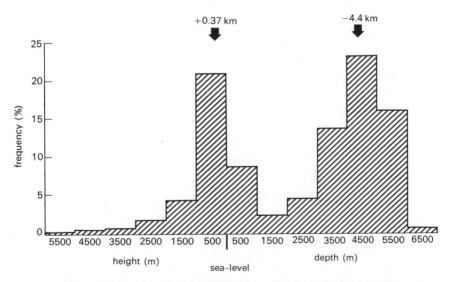

Figure 1.3 The average elevation of the land above sea level, and the average depth of the oceans (modified from McLellan, 1965, and Weyl, 1970).

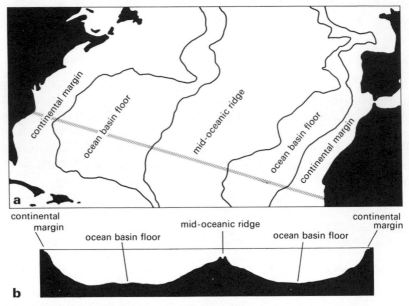

Figure 1.4 The three major levels or physiographic divisions that are distinguishable in most oceans: the continental margins, the ocean basin floor and the mid-oceanic ridge. (*a*) Southern North Atlantic (part of North America is seen on the left, and parts of Spain and North Africa on the right). (*b*) A vertical transect across the Atlantic along the dotted line in the upper diagram (modified from Heezen, Tharp and Ewing, 1959).

shore to 130 to 200 m depth. The widths of continental shelves vary considerably; for instance, there is none around parts of Australia, while the whole of the North Sea is continental shelf. The shelves have an average width of about 50 km, and represent about 8% of the world's seas in area. Although small in area, man knows them best. They have been explored by research vessels, manned submersibles, and SCUBA divers, have often been heavily fished and have enough oil, gas and coal beneath them to be tapped by commercial operators.

(ii) *The ocean basin floor*

From 200 m to between 1500 and 4000 m, the continental shelf steepens into the *continental slope*, at about 3 to 6° to the horizontal, and from there to about 4000–5000 m flattens on to the *continental rise* at about 0.1 to 1°(Figure 1.4). The continental slopes and rises are cut by many *submarine canyons*. Alluvial fans of sediment spread out from the seaward end of these

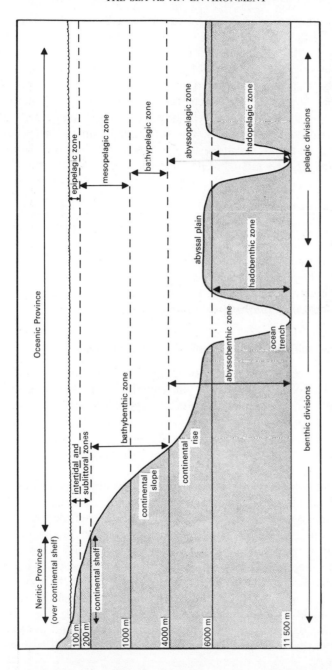

Figure 1.5 Cross section of part of an idealised ocean, with the various zones of the benthic and pelagic divisions indicated.

canyons, probably carried down by turbidity currents (Chapter 6); these sediments give the *abyssal plains* a very flat surface. From the edges of the continental rise the completely flat abyssal plains stretch for hundreds of miles with a slope of less than 0.1°, sometimes punctuated by shallow abyssal hills. They are interrupted by deep ocean trenches such as the 11 000 m deep Marianas Trench near the Philippine Islands in the Pacific, by various volcanic and coral reef islands, and by submerged ridges, volcanoes, *seamounts*, and *guyots*. The submerged seamounts are circular or elliptical from above, rise at least 1000 m from the abyssal plain and are volcanic. Guyots are similar but flat-topped.

(iii) *The mid-oceanic ridge*

The submerged ridges and volcanic islands are often aligned along the centre of the major oceans (Figure 1.3); the mid-Atlantic Ridge running roughly north/south along the centre of the Atlantic Ocean is an example. They represent lines of earthquake and volcanic activity in the earth's crust, where new crust is being formed or where plates of the earth's crust are pushing against one another (see Chapter 5).

Groupings of animals and plants, major ecological zones

Although the oceans cover most of the earth, they contain many fewer species than exist in fresh water or on land. There are about one million described animal species on earth, but only 160 000 or 16% of these live in the oceans, and of these only 2% (*c.* 3200) live in mid-water—the remainder live on or in the sea bed. Lack of diversity, however, is made up by huge numbers of some groups—for example, the pelagic copepods and benthic molluscs.

Animals and plants that float or swim in the sea are called *pelagic*, while those that live on or in the sea bed are called *benthic*. Pelagic animals and plants that inhabit the top few centimetres of the sea are called *neuston* and are microscopic. Slightly larger pelagic animals and plants that float passively or swim weakly are termed *zooplankton* and *phytoplankton* respectively. Larger pelagic animals that can swim well, such as many fish and cephalopods, are called *nekton*. The exact size limits of these groups are rather vague (Chapter 4). Neustonic organisms are generally less than 500 μm, zooplankton and phytoplankton about 500 μm to 2 or 3 cm, and nekton from 2 or 3 cm upwards. Phytoplankton are found in the upper sunlit zone of the ocean (0–300 m); zooplankton are also found there but extend to great depths; nekton are widely distributed throughout the depths of the ocean.

Benthic organisms include such groups as the intertidal seaweeds, many crustacea, burrowing bivalves, echinoderms and polychaetes, and live in or on rock or sediments of mud and sand. They are found from the shallowest to the deepest parts of the ocean.

The environments in which pelagic and benthic organisms live are classified in a similar way (Figure 1.5). Benthic organisms live in the *intertidal zone* between tidemarks, in the *sublittoral zone* from low tide level to the edge of the continental shelf, and in the progressively deeper *bathybenthic, abyssobenthic,* and *hadobenthic zones*. The intertidal and sublittoral zones are sometimes grouped into the *littoral zone*, but this is confusing and will not be used in this book.

Pelagic animals live over the continental shelves in the *Neritic Province* or over the deep oceans in the *Oceanic Province* (Figure 1.5). Vertically their environment is divided as follows. The upper sunlit *epipelagic zone* extends from the surface to about 100 m, and has strong light and temperature gradients which often vary with the seasons. The *mesopelagic zone*, 100 to 1000 m, is dim or totally dark, has a fairly even temperature, and sometimes an oxygen minimum and nitrate and phosphate maximum layer (see Chapter 3). Below this are the *bathypelagic, abyssopelagic,* and *hadopelagic zones* extending to the greatest depths of the ocean. The distinctions between the different zones are not exact, and experts disagree on some of the definitions.

The benthic and pelagic environments can be grouped together to give four major ecological zones: *intertidal* and *sublittoral* (equals littoral), *bathyal, abyssal,* and *hadal* (Table 1.2). In general, temperatures decrease and fluctuate less with increased depths, pressure increases, and light decreases. The abyssal zone has the largest surface area, and the hadal the smallest surface area.

Table 1.2 Major ecological zones of the world's oceans

Characteristic	Littoral and sublittoral	Bathyal	Abyssal	Hadal
Area of sea bed (%)	8	16	76	1
Depth (metres)	0–200	200–4000	4000–11 500	6000–11 500
Pressure (atmosphere)	1–21	21–401	401–1151	601–1151
Temperature (°C)	25–5	15–5	< 5	< 3.5
Light	bright–dim	dim–dark	total darkness	total darkness

CHAPTER TWO

OCEANIC CIRCULATION AND WATER MOVEMENT

Energy gain and loss, earth, oceans

Energy from the sun in the form of electromagnetic radiation reaches the earth as shortwave radiation in the ultraviolet, visible and infrared, with a peak at a wavelength of 0.5 μm (Figure 2.1). Assuming no atmosphere, sea and land surfaces exposed perpendicularly to the sun's energy would receive about 1.36 kW m^{-2} (this figure is called the solar constant). They receive much less because scattering, absorption, and reflection takes place in the atmosphere (Table 2.1). On clear days 80% reaches the ocean's surface, but on cloudy days the figure is only 25%. At a water depth of 10 metres these percentages are reduced to 8% and 2.5% in oceanic waters and 0.4% and 0.125% in coastal waters.

The temperature of the earth and oceans remains roughly constant on a scale of years and decades because the energy absorbed from the sun is balanced by re-emission back into space from the earth. It is re-emitted at a

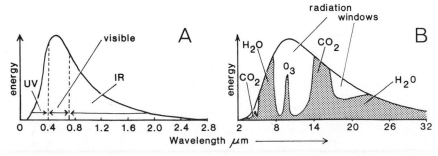

Figure 2.1 *A*: Distribution of radiation intensity with wavelength for a black body having a surface temperature of 6000 K, representing the sun. *B*: distribution of radiation intensity with wavelength for a black body having a surface temperature of 285 K, representing the earth. Absorption of radiation by water vapour, carbon dioxide, and ozone in the atmosphere is shown schematically by shading (after Harvey, 1976).

Table 2.1 Absorption of radiation in the atmosphere and ocean (modified from Harvey, 1976).

	Clear sky	Cloudy sky
Incoming radiation ($c.\,1.36\,kW.m^{-2}$)	100%	100%
Scattering back to space by air and small particles	7%	7%
Absorption in upper atmosphere (mainly O_3)	3%	3%
Absorption in lower atmosphere (mainly water vapour)	10%	10%
Clouds		
reflection to space	—	45%
absorption	—	10%
Reaches ocean surface	80%	25%
Reaches 10 m		
Oceanic waters	8%	2.5%
Coastal waters	0.4%	0.125%

longer wavelength than it is received ($c.\,2$–$32\,\mu m$), because the temperature of the earth is much lower than that of the sun. This follows from Stefan's law:

$$E = \sigma T^4$$

where E = radiation emitted per unit surface area from an equivalent black body to the sun or earth, σ = a constant, and T = absolute surface temperature of a black body equivalent to the sun (6000 K) or earth (250 K). Lower temperatures produce less energy which means a longer wavelength. However, the earth's average temperature is not 250 K ($-23\,°C$) but $c.\,285$ K ($12\,°C$) because some of the energy emitted from the land and sea is absorbed by water vapour, carbon dioxide and ozone in the atmosphere. This means that energy can only leave the earth through a major radiation window at 8 to $12\,\mu m$ and a smaller one at $c.\,16$ to $22\,\mu m$ (Figure 2.1). At other wavelengths it is completely absorbed.

Energy is lost from the ocean to the atmosphere by evaporation (54%), radiation (41%) and conduction (5%). Considerable energy is needed to evaporate water from the sea surface, because the latent heat of evaporation of water is large. Long-wavelength energy in the infrared is radiated from the sea surface to the atmosphere and space. Direct transfer of energy from the ocean to the atmosphere takes place by conduction because the ocean is usually slightly warmer than overlying air. Although there are large seasonal and geographical differences, this overall oceanic heat budget applies between about 70° N and 70° S.

Variations in heat transfer between ocean and atmosphere, seasons

The amout of both light and heat radiation received by the oceans from the sun varies with the seasons as one moves away from the equator. The maximum energy received in summer at the top of the atmosphere actually increases slightly towards the poles, but the minimum energy received in winter falls dramatically (Figure 2.2). These changes are the cause of the seasons in temperate climates. Energy is transferred between the atmosphere and the sea at the air/water interface. In northern temperate climates the air is usually a few degrees warmer than the sea from April to October, and a few degrees colder from November to March, and so heat is transferred from air to water from April to October and from water to air from November to March. Heat is transferred by conduction across the

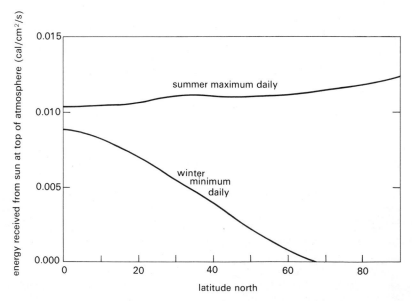

Figure 2.2 Energy received from the sun at the top of the earth's atmosphere, in relation to latitude in the northern hemisphere (modified from Weyl, 1970).

air/water interface (sensible heat), by the evaporation of water (latent heat of evaporation), and by melting ice (latent heat of freezing).

Adiabatic temperature changes

When air is compressed, work is done on it and so its temperature rises. For example, a bicycle pump warms up as air is pumped from it into a bicycle tyre. Air is compressed fairly easily, but water is only slightly compressed with an increase in pressure. However, at great depths, below about 4000 m (400 atm) the pressure is enough to raise the water temperature by about 1 °C. The temperature increase occurs without application or transfer of heat from surrounding water masses, and is called *adiabatic*. If the body of water were to be raised to the surface, it would expand and so cool a degree or so; this cooler temperature is called its *potential temperature*.

Water cycling, oceans, land, atmosphere

Water continuously moves between the atmosphere, the oceans, fresh waters and the land in what is called a *hydrological cycle*. As it does so it

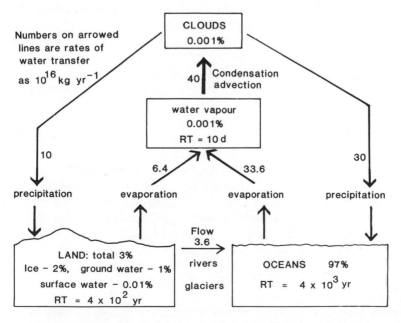

Figure 2.3 The hydrological cycle (after Harvey, 1976). RT, residence time.

changes from solid (ice) to liquid to gas and moves from place to place (Harvey, 1976; Press and Siever, 1982; Strahler and Strahler, 1983) (Figure 2.3). About 97% of the total global water is stored in the oceans, 3% in or on the land and fresh waters, and 0.002% in the atmosphere (water vapour + clouds). The residence time of water in the sea is 4000 years, on the land is 400 years, and in the atmosphere is only 10 days. The residence time is the average time between entering and leaving a compartment of the cycle. A small amount of juvenile water is also added to the global total from within the earth by geochemical processes such as volcanoes.

The processes of evaporation, condensation, and precipitation are critical to the transfer of water to and from the oceans and land. Evaporation of water from the surface of the sea and land into the atmosphere produces a vapour pressure. The vapour pressure of an air mass is the pressure of the gaseous water molecules in it. Condensation of water vapour at the surfaces of the sea and land reduces the vapour pressure. When the air is saturated with water, the vapour pressure is termed the saturation vapour pressure. This pressure increases rapidly and non-linearly with temperature and is about 6 mb (millibars) at 0 °C, 12.5 mb at 10 °C, c. 23 mb at 20 °C, and 42 mb at 30 °C.

Gaseous water molecules in the atmosphere can condense on to water, ice or land surfaces, or on to salt or dust particles in the atmosphere called condensation nuclei. Condensation usually happens when the relative humidity just exceeds 100%. Below 100% condensation takes place on to hygroscopic nuclei which are usually soluble salt particles or industrial pollutants. The progressive saturation of air that leads to condensation usually occurs because the air is cooled.

Precipitation is the formation of water droplets following condensation. The water droplets in clouds are very small. They have a radius of 10 μm and fall slowly at about 0.01 ms^{-1}. Raindrops that reach the surface of the sea and land are much larger (c. 1000 μm) and fall at metres per second. The small droplets must therefore coalesce in some way in clouds. It is not known exactly how this happens, but it probably occurs by random collisions and by coalescence onto rapidly falling ice crystals from supercooled water, the Bergeron–Findeisen theory. However this latter effect is likely to happen only in the middle and high latitudes.

Salinity

The evaporation of water from the sea surface increases the salinity of the sea as well as transferring latent heat from the ocean to the atmosphere.

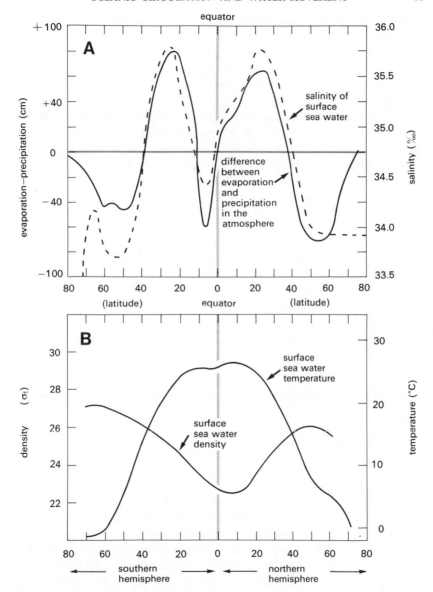

Figure 2.4 *A*. Close direct relationship between the variation in salinity of surface waters of the oceans, and the difference between evaporation and precipitation in the atmosphere. The two variables are plotted against latitude on the x-axis (modified from Groen, 1967). *B*. Inverse relationship between the density and temperature of surface sea water. The two variables are plotted against latitude on the x-axis (modified from Pickard, 1975).

Once water sinks below the surface, salinity cannot be increased by evaporation. For the salinity of the seas as a whole to remain constant at 35 parts of salt per 1000 of water, the increase must be balanced by a reduction elsewhere. Melting of polar ice, precipitation (rain and snow), and fresh water from rivers all reduce the ocean's salinity. The variation in salinity in surface waters is very similar to the balance between evaporation and precipitation, and appears to be largely controlled by this balance in most areas (Figure 2.4A). In polar and temperate regions between 40° and 70° N and S (and surprisingly also at the equator) precipitation exceeds evaporation and surface salinity is low. Between about 10° and 40° N and S, evaporation exceeds precipitation and salinity is high (Figure 2.4A).

Density of sea water

The density of sea water increases with its salt concentration and pressure, and decreases with temperature. One part per 1000 of salt increases the density about 0.8 parts per 1000; therefore, 35 parts per 1000 of salt (normal sea water) gives a density of 1.028 g/cm^3. This is normally measured at 0 °C and at a pressure of one bar. Density is expressed by oceanographers as 'sigma t' (σ_t). 1.028 g/cm^3 = a sigma t of 28. The average density of surface waters is low at the equator and rises in higher latitudes, and is inversely related to temperature (Figure 2.4B).

Temperature and salinity measurements, reversing bottles, TSD probes

Measurements of the sea's temperature and salinity at the same time and place enables one to calculate its density using tables of density (σ_t) for a range of temperatures and salinities. Temperature and salinity are measured by *Nansen reversing bottles* or by *temperature–salinity–depth* (TSD) probes.

Nansen reversing bottles are usually plastic or metal and open at each end, and carry highly accurate thermometers. They are attached to a wire lowered from the side of a stationary research vessel at 50 m, 100 m, 250 m and so on, to give a representative coverage of the water column. A cylindrical metal messenger slides down the wire and triggers the first bottle to turn over and shut its ends, thus enclosing a sample of water from 50 m. This releases another messenger which slides down the wire to trigger the next bottle and so on. When all the bottles have been triggered, the wire is hauled in, the thermometers read, and the water run off for future analyses of salinity, oxygen and nutrients. Salinity can be measured by titration, or as electrical conductivity using a salinometer. The latter is easy and quick, particularly on board ship, and is very accurate. It can detect differences of less than 0.01‰, equivalent to the increased salinity caused by evaporation when sea water is poured from one container into another. For accuracy, the salinity of an unknown sea water is usually compared with standard sea water, supplied by IAPSO Standard Sea Water Service at the Institute of Oceanographic Sciences, Godalming, Surrey, UK.

Temperature–salinity–depth electronic probes record data continuously as they are lowered through the water column. Probes such as these have given us a knowledge of the fine microstructure of the ocean that cannot be obtained from the stepwise data available from reversing bottles. The 1–2 m high cylindrical probe, to which water sampling devices can be attached, is lowered by cable from a stationary research vessel, and data are transmitted along the cable to the ship.

Seasonal and geographical changes in temperature. The thermocline

The temperature of the sea ranges from 2 °C to | 30 °C but the overall average is + 3.5 °C, so most of the world's oceans are cold. In polar seas

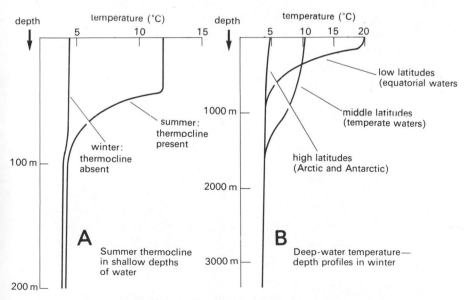

Figure 2.5 Changes in temperature with depth. *A*. Summer thermocline in shallow depths of water. *B*. Winter temperature profiles at deeper depths. The depth scales are only approximate as there is much variation (modified from Herring and Clarke, 1971, and Pickard, 1975).

temperatures are low throughout the year and roughly constant with depth (Figure 2.5*B*). In middle latitudes in summer, the temperature of the sea surface is 10–18 °C. Below the surface, temperature remains fairly constant with increasing depth, until the water suddenly becomes cooler over a short vertical depth. This sudden change in temperature with depth is called the *thermocline*. In tropical waters the thermocline is usually close to the surface and very sharp. In temperate climates the thermocline develops from a uniform temperature–depth distribution in early spring and then decays again in autumn (Figure 2.5*A*), and the abundance of phyto- and zoo-plankton in the sea is closely linked with its seasonal development and decay (Chapter 4). It is also worth noting the exact position and sharpness of the thermocline can change hourly—with cloud cover for example.

Climatic ocean regions

Elliot (1960) has divided the oceans of the world into a number of climatic regions based on their temperature, salinity, evaporation and precipitation.

(1) Low-latitude tropical oceanic waters between 30° N and S are warm, have winter temperatures of 20 to 25 °C, and have an annual temperature variation of less than 5 °C. Evaporation usually exceeds precipitation, and so salinity is often high, 35–37‰. Many currents are westerly, since the trade winds blow in these regions.

(2) Mid-latitude oceanic waters lying between the tropics and the polar regions, that is, between about 30° and 60° N or S, are cooler, have winter temperatures between 20° and 5 °C, and an annual variation of about 10° C. Precipitation is slightly greater than evaporation, and salinity is equal to the ocean average, 35‰. Surface temperatures and salinities are very variable seasonally and from place to place. The major currents move eastwards under the influence of the westerlies. Moving depressions with rain and snow are common.

(3) High-latitude oceanic waters constituting most of the Arctic and Antarctic Oceans are covered for much of the year by ice, and have temperatures of about 0 °C for most of the year. A temperatures are usually below 0 °C, but can fluctuate 40° seasonally. Precipitation is low but exceeds evaporation, and salinities are low, 28–32‰. Winds and currents circulate clockwise around the poles, particularly in the Antarctic.

Atmospheric circulation

The tropical regions of the earth receive more solar radiation per unit area than the polar regions because the latter receive energy at a low angle, which obviously reduces the energy received per square metre. On the other hand, both tropical and polar regions radiate about the same amount of long-wavelength radiation back into space. As a result, air over the tropics becomes much hotter than air over the poles. This causes a three-cell circulation, and alternating regions of high and low pressure in the atmosphere (Figure 2.6). Hot air at the equator expands and rises, and pressure is therefore low. During this process the air loses heat. From the equator the mass of air moves northwards and southwards. As it moves at a constant height above the earth, it becomes compressed because of the earth's curvature; for example, if it reached the North Pole it would have to be compressed into a very small volume. (In order to understand this process, it may help to imagine an equilateral triangle on the surface of the earth, whose base lies along the equator. Any area of the triangle near the base would become compressed by the sides of the triangle if it were moved towards the apex.) Between latitudes 20 and 35° N or S, the air compressed in this way is dense enough to sink. When it meets the ocean surface it spreads out north and south again; this is the high-pressure subtropical calm belt on either side of the equator.

North-moving low-altitude air and south-moving low-altitude air do not move directly north and south. They are deflected towards the right in the

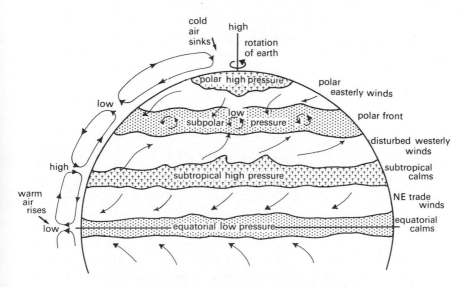

Figure 2.6 Wind circulation in the northern hemisphere.

northern hemisphere and towards the left in the southern hemisphere. As a result, in the northern hemisphere the south-moving low-altitude wind is deflected towards the west to give the *north-east trade winds*, and the north-moving low-altitude wind is deflected towards the east to give the *westerlies* (winds are named by the direction from which they blow). The NE trade winds are fairly constant while the westerlies are more variable. Cold air descends at the poles, producing a high-pressure zone; it then moves southwards near the earth's surface and is deflected towards the west. The resultant polar easterly winds meet the variable westerly winds face to face and cause disturbed weather at the polar front. In this way three vertical cells of rotating air masses are formed in the northern and southern hemispheres (Figure 2.6).

Coriolis force

The deflection to the right in the northern hemisphere and to the left in the southern hemisphere is caused by the Coriolis force which may be explained as follows. The earth spins from west to east about its north-south axis. The rotational velocity, or speed, of a point on the earth's surface is therefore greatest at the equator and decreases towards zero at the poles. Imagine the earth seen from the North Pole. The distance travelled by a point on the earth's surface at the equator is 40 000 kilometres per day or 1670 km h^{-1}, and at latitude $60° \text{ N}$ or S, where the radius of the circle of latitude is half that at the equator, it is 20 000 km per day or 835 km h^{-1}. At the North or South Pole the point has no speed since it is revolving on its own axis. Now consider a mass of air or water (which, of course, can move in relation to the earth below it) at the equator and travelling at the same speed as the earth below it (1670 km h^{-1}). If the mass moves northwards or southwards away from the equator, it will tend to retain its initial velocity of 1670 km h^{-1}, since it is able to slip over the earth's surface, while the earth beneath it at its new position will be moving more slowly (e.g. at 835 km h^{-1} at $60° \text{ N}$ or S). In other words as the mass of air or water moves northwards or southwards away from the equator it will be moving eastwards (in the direction of the earth's spin) progressively more quickly than the earth beneath it. This difference in speed between the mass moving over the surface of the earth and the speed of the earth's surface at the same point can be regarded as a deflection of the mass towards the right of its initial northerly direction of movement in the northern hemisphere, and as a deflection towards the left of its initial southerly direction of movement in the southern hemisphere.

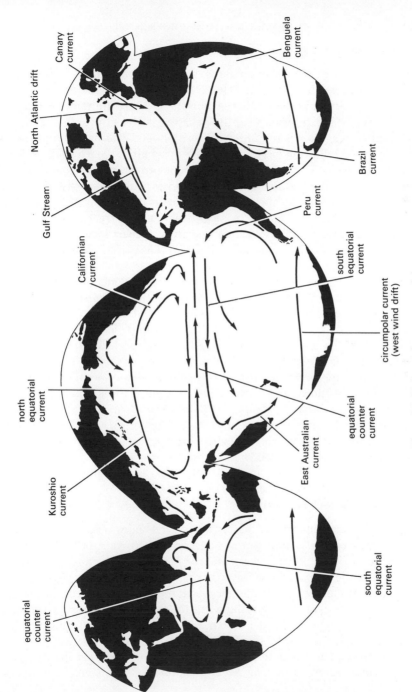

Figure 2.7 Major surface currents of the oceans.

Major ocean currents

The major ocean currents are similar in most oceans (Figure 2.7). They flow in a roughly circular pattern or gyre, clockwise in the northern hemisphere and anti-clockwise in the southern hemisphere. There are sometimes areas of relatively stagnant water near their centres, such as the Sargasso Sea in the North Atlantic where *Sargassum* weed floats. The pattern of flow is caused by the following mechanisms. Firstly, winds at the sea surface move the surface water. The influence of the wind is progressively reduced as one moves away from the water surface into deeper water. Secondly, the shape of the ocean basins and surrounding continents directs the ocean currents in a circular motion. Thirdly, tropical waters are warmer and less dense than polar waters and are therefore slightly higher; hence water has a tendency to flow from the tropics to the poles. The Coriolis force acts on the winds and on the ocean currents produced by these mechanisms, and water currents are hence deflected to the right of the wind direction in the northern hemisphere and to the left in the southern hemisphere.

Ocean currents are named by the direction in which they are flowing. In contrast, winds are named by the direction from which they have blown (Figure 2.6). The most important winds producing ocean currents at the sea surface are the *North-East Trades* and the *Westerlies* in the northern hemisphere, and the *South-East Trades* and the *Westerlies* in the southern hemisphere. In addition, the *circum-polar current* in the southern hemisphere is driven by westerly winds (the *West Wind Drift*) as a ring of water around the earth at between 45° and 60° S. There is no equivalent circum-polar current in the northern hemisphere because there are many land masses between 45° and 60° N.

Surface and subsurface currents

In the northern hemisphere, the currents of the major ocean gyres are fast and narrow on the eastern margins of Asia and North America; these are the *Kuroshio* in the Pacific Ocean and the *Florida Current* and *Gulf Stream* in the Atlantic Ocean. These fast western boundary currents change their local position unpredictably. On the western margins of the continents the currents are slow and broad; these are the *North Pacific* and *California Currents* in the North Pacific Ocean and the *North Atlantic* and *Canary Currents* in the North Atlantic. Where the northern hemisphere gyres come close to the equator in all three major oceans, Indian, Pacific and Atlantic, they form a westward-moving *North Equatorial Current*. In the southern

Table 2.2 Water transport by the major surface currents of the oceans (Weyl, 1970).

Currents	Transport rate $(m^3 \times 10^6\, s^{-1})$
Antarctic circum–polar current	200–150
Gulf Stream–Cape Hatteras	100
Kuroshio current	50
Gulf Stream–Florida Straits	25
All the world's rivers	1

hemisphere, westward-moving *South Equatorial Currents* form in the same way. The South and North Equatorial Currents are separated by a current flowing in the opposite direction, the *Equatorial Counter-Current*. Recently an equatorial undercurrent, the *Cromwell Current*, has been discovered running eastwards exactly at the equator under the westward-flowing Equatorial Currents. It flows at depths of 40 to 100 m, is 300 km wide, 200 m deep, and flows at up to $5\,km\,h^{-1}$.

Weyl (1970) has calculated the rates of water transport by the major currents at the ocean's surface. They are remarkably high, in some cases more than a hundred times that of all the world's rivers together (Table 2.2),

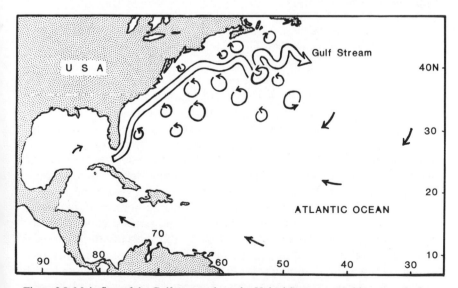

Figure 2.8 Main flow of the Gulf stream along the United States coast with mesoscale rings spawned from the main current. Based partly on satellite imagery (after Couper, 1983).

and must therefore keep the surface waters of the oceans well mixed. The major surface currents also affect continental climates; for example, Britain is north of Newfoundland, but has a warmer climate because the warm North Atlantic Drift flows near it.

Infra-red satellite imagery and satellite tracking of free drifting buoys have shown that many of the major current flows are complex. The Gulf Stream has a number of meanders and eddies; surface water criss-crosses the main current, and self-enclosed rings of water ranging in diameter from 20 to 150 km break away from the main current (Figure 2.8). Three clockwise rings usually form north of the Gulf Stream and 8–14 anticlockwise ones to the south. Both sets travel against the main north-easterly direction and appear to feel the bottom since they sometimes circulate around and over sea mounts.

Current measurement

The speed and direction of the major surface and near-surface currents are measured by floats at or just below the water surface, by current meters, and also by measuring the temperature and salinity of the currents. Temperature and salinity enable one to calculate pressure gradients, and hence to plot the movement of water down the gradient. Surface buoys are now in use which transmit information either directly to the shore or to a satellite. Floats are released from research vessels, and temperature and salinity measured while the vessel is stationary by lowering a temperature-salinity-depth (TSD) probe to which is often attached a current meter. The bathythermograph, which is a metre-long metal cylinder towed at up to 3 knots by a research vessel, measures temperature and depth by a sensor recording onto a small smoked glass plate. An expendable model has recently been developed, which allows a research vessel to continue cruising at full speed (8–12 knots) and so to save time and money (large research vessels may cost £1000 to £9000 per day to operate). Sophisticated pop-up systems carrying an array of electronic recording devices (temperature, current, salinity, depth) are now in use by a number of marine research laboratories. They are weighted to sit on the bottom, and to record continuously. To retrieve them a research vessel sends a coded acoustic signal which detonates a small explosive charge; this releases the equipment from the weights and allows it to float to the surface under its own buoyancy (Figure 2.9). Biological indicators are also sometimes used, such as the appearance or disappearance of warm or cold-water planktonic species (see Chapter 4).

A

Neutrally Buoyant Floats

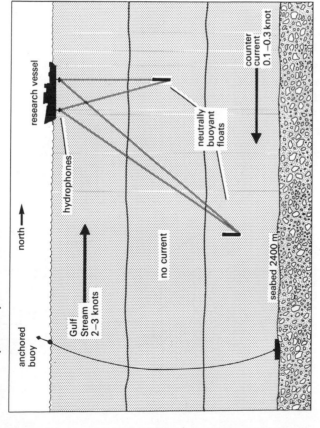

B Pop-up Recorder

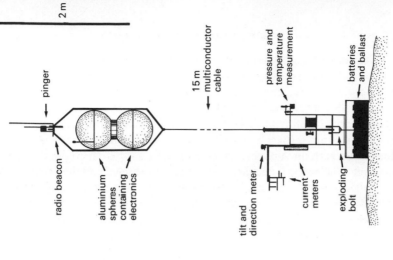

Figure 2.9 Measuring ocean currents. *A*. Neutrally buoyant floats at different depths are interrogated by pinger from a research vessel. The movement of the floats is measured in relation to an anchored buoy (modified from Herring and Clarke, 1971). *B*. Pop-up temperature, current and pressure recorder sitting on the ocean bed at 1000 m, operated by the Scripps Institute of Oceanography (Snodgrass, 1968).

The currents and circulation of the deep oceans are more difficult to study and so less is known of them. They are mapped from salinity and temperature data, from the movements of neutrally buoyant floats (Figure 2.9), by the use of such equipment as *Carruther's Pisa*, and by studying the bending of stalked benthic invertebrates and ripples on the sea bottom (Figure 6.16). The *Pisa* is a small bottle half-filled with gelatine and topped up with oil. A magnet is suspended by a thread through the oil to the gelatine. The bottle is heated until the gelatine melts, and then released with a small weight or anchor. If heated to the right temperature, the gelatine will not solidify until the bottle reaches the sea bottom. The angle of the thread through the solidified gelatine to the vertical is then proportional to the current speed, and the magnet shows current direction.

Vertical circulation in the Antarctic

The movement of water in the deep oceans is best understood near the Antarctic Continent (Figure 2.10). Melting ice near the edge of the Antarctic Continent produces low-salinity water that mixes with the southward-moving Antarctic Deep Current. From this mixture is produced firstly the north-flowing *Antarctic Surface Current* which then sinks to become the *Antarctic Intermediate Current* or *water*, and secondly the

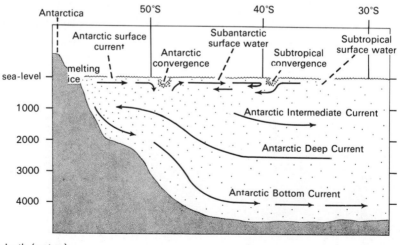

Figure 2.10 Vertical circulation near the Antarctic Continent (modified from Hill, vol. 2, 1963).

Antarctic Bottom Current or *water* which is relatively saline but very cold and which flows northwards along the ocean bottom. The Antarctic Bottom Current can sometimes be traced well into the northern hemisphere in the Atlantic Ocean.

The *Antarctic Deep Current* originates from warm high-salinity subsurface water that flows southwards from the Pacific, Atlantic, and Indian Oceans. This water becomes progressively deeper and less saline until at between 1500 and 3000 m it is termed the *Antarctic Deep Current*. At about 50° S the current moves upward between the Antarctic Bottom Current and the Antarctic Intermediate Current (Figure 2.10) and eventually mixes with low-salinity water formed from melting ice at the edge of the Antarctic Continent.

The exact disposition of the currents produced around the continent of Antarctica depends on the local bottom topography and also on water density, which in turn depends on salinity and temperature. (For example, it is obvious that more ice will melt to produce low-salinity water at the edge of the continent in summer than in winter.) Higher density water will sink below lower density water. High salinity and low temperature increase the density of water, and low salinity and high temperature decrease it. The salinity and temperature differences need only be very small to produce the density differences which will separate the currents. Normal sea water contains 35 parts per thousand of salt (‰), that is 35 g/1. The Antarctic Intermediate Current has a low salinity of 34.10‰, and the Antarctic Deep Current a higher salinity of 34.75‰. The Antarctic Deep Current is relatively warm at $+2.8\,°C$, the Antarctic Bottom Current is cold at $-0.8\,°C$, while the Antarctic Intermediate Current has a temperature of $+2.0\,°C$.

Two convergences of surface currents occur near the Antarctic Continent: the *Antarctic Convergence* and the *Subtropical Convergence* (Figure 2.10). The position of each is clearly marked by different salinities, by different temperatures, and by different planktonic species.

The vertical circulation of water in the Antarctic is part of the more general global picture of deep-water circulation, which is now described.

Global movement of water masses, deep water circulation

There are two main sources of deep water in the oceans (Figure 2.11). Water sinks from the surface in the Arctic and Antarctic because it is cold and has a high salinity, and thus is dense. Water in the Norwegian Sea and off Greenland sinks to form the *North Atlantic Deep Water* which moves

North Atlantic Deep Water   Pacific Deep Water

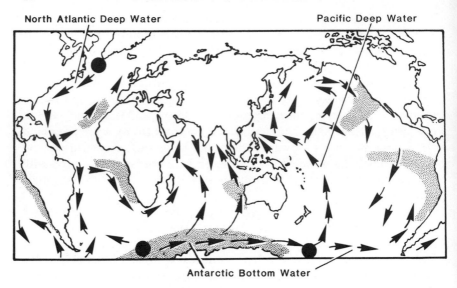

Antarctic Bottom Water

Figure 2.11 Movement of oceanic bottom water. ●: Main sources of bottom water are in the Arctic and Antarctic. Shading represents major areas of upwelling. Velocities: western margins of ocean basins $\leqslant 20\,\mathrm{cm\,s^{-1}}$; elsewhere $\leqslant 0.5\,\mathrm{cm\,s^{-1}}$ (after Couper, 1983).

south-west, and water in the Antarctic Ocean (Weddell Sea) sinks to form *Antarctic Bottom Water*. The former is more important than the latter. *Pacific Deep Water* is probably formed by a 50:50 mixture of *North Atlantic Deep Water* and *Antarctic Bottom Water*. These bottom waters move at $c.\,0.2$ to $20\,\mathrm{cm\,s^{-1}}$. They rise to the surface again at the major upwelling areas on the west coasts of continents and in the Antarctic (Figure 2.11).

Much of the evidence for this and for the movement of other major water masses such as that of the *Mediterranean Outflow Water* comes from two methods. The first is the linear relation between salinity and the δO^{18} of different water masses.

$$\delta O^{18} = \frac{(^{18}O/^{16}O)\text{sample} - (^{18}O/^{16}O)\text{SMOW} \times 1000}{(^{18}O/^{16}O)\text{SMOW}}$$

where SMOW = *Standard Mean Ocean Water*, and $\times 1000$ expresses δO^{18} as parts per thousand deviations in the $^{18}O/^{16}O$ ratio of the sample from that of SMOW. δO^{18} ranges from $-1.0‰$ to $+0.5‰$ (i.e. per mil) for the larger oceanic water masses. Each water mass has a characteristic δO^{18}/salinity ratio. Figure 2.12A shows that North Atlantic deep water is on the line for North Atlantic Surface Waters and that Antarctic Bottom

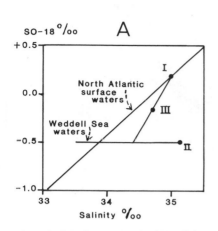
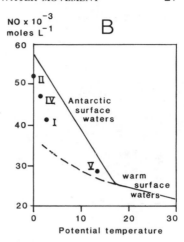

Figure 2.12 Indirect methods of identifying and hence following the movements of deep water in the oceans. I, North Atlantic Deep water; II, Antarctic Bottom water; III, Pacific Deep water; IV, North Pacific Bottom water; V, Mediterranean Sea outflow waters. A, relationship between salinity (‰), δO^{18}(‰), B, relationship between potential temperature and NO $\times 10^{-3}$ moles L^{-1} (after Broecker, 1974).

Water is on an extension of the Weddell Sea Line. Pacific deep water falls almost exactly halfway between the two.

The second method uses the inverse relationship between the potential temperature and the NO value of different water masses. Each water mass has a characteristic NO/potential temperature ratio. The NO of a water mass is the sum $(7 \times NO_3^-) + O_2$ in molar concentrations. The NO value of a given water mass is constant at a given temperature despite changes that occur in dissolved O_2 and NO_3^- as the water mass moves into deeper waters, and despite changes in respiration by microorganisms and plankton. The sum remains constant because for each 7 moles of oxygen used in respiration—and hence removed from sea water—one mole of nitrate is released into sea water. In solution therefore the sum $(7 \times NO_3^-) + O_2$ in moles stays the same whether respiration rises or falls. This equation can only be approximate, however, since it ignores removal of nitrate by nitrate-reducing bacteria and by phytoplankton during photosynthesis.

These two methods allow major water masses deep in the oceans to be identified and then tracked. The δO^{18} and salinity (Figure 2.12A) or the NO and potential temperature (Figure 2.12B) of a particular water mass are measured, and then related to values of known water masses on the two graphs.

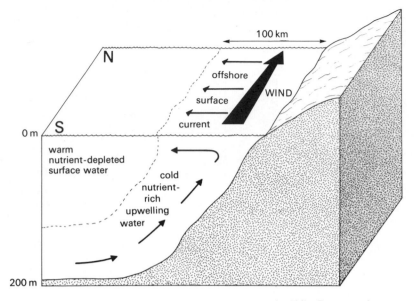

Figure 2.13 Upwelling of deep cold nutrient-rich water along the Chile–Peru coast (west coast of South America).

Upwelling

Along the western coasts of North and South America, and the western coast of Africa, deep cold water is brought to the surface during the summer by wind action (Figure 2.13). On the Oregon coast, USA, for example, northerly winds blow along the coast, and the coastal surface water is moved not southwards but away from the coast by the Coriolis force. Deep water then wells up to take its place. The upwelling water in these areas has a high nutrient content and so produces high productivity and good fisheries, such as the anchoveta fishery off the coast of Peru.

Vertical and horizontal mixing in the thermocline

The thermocline has a density gradient across it and hence is a major barrier to vertical mixing in the surface layers of the ocean. Lateral and vertical mixing in the thermocline are measured by vertical profiles of salinity and temperature and radioactive markers (^{137}Cs, ^{90}Sr, ^{14}C, ^{3}H). These radioactive elements were released into the atmosphere by the thermonuclear tests of 1954–64, progressively entered the ocean, and are now useful artificial markers.

In the equatorial Atlantic little radioactivity is found below 60 m, thus implying that vertical mixing only occurs to this depth. Here the thermocline has a sharp density gradient and is close to the surface. In contrast, in the North Atlantic at 35° to 45° N radioactivity is found at 700 m and so vertical mixing is deeper. Here the thermocline is deeper and less distinct thus providing less of a vertical barrier.

Water in the thermocline region spreads horizontally over long distances by laminae of water moving along horizons of constant density from which there is little vertical mixing. For example, in the North Pacific, ^{3}H and ^{137}Cs maxima at 100 m exactly match a salinity minimum at this depth which is caused by sinking and lateral spreading of water from north of Japan. The maxima have been traced 1600 km across the Pacific over ten years which suggests that little vertical mixing can have occurred over this period.

Vertical mixing in deeper waters

Vertical mixing between 1000 m and 4000 m has been studied in the Pacific using salinity/depth and potential temperature/depth profiles, and gradients of isotopes of carbon (^{14}C), radium (^{226}Ra) and silicon (^{32}Si). Low salinity water at 1000 m starts as Antarctic Intermediate Water and moves northwards at that level. It progressively mixes with North Atlantic Deep Water and Antarctic Bottom Water lying between 1000 m and 4000 m, at vertical mixing rates of 2 to 5 m yr^{-1} and vertical diffusivity of coefficients of 1.5 cm^2 s^{-1}. The mixing is probably caused by uniform upward flow of deep water (advection) and random eddy diffusion (diffusion mixing).

Vertical mixing near the sea bed

Vertical mixing near the sea bed can be measured using an isotope of radon (^{222}Rn). ^{222}Rn decays from ^{226}Ra within and above the sediment and has a half-life of four days. Its concentration thus decreases away from the sediment/water interface, and the shape of this concentration gradient gives estimates of mixing within the bottom 100 m. Coefficients of vertical eddy diffusivity calculated from this curve are 2 to 200 cm^2 s^{-1} which are many orders of magnitude greater than coefficients of molecular diffusion (0.5 to 16×10^{-5} cm^2 s^{-1}). So vertical mixing by currents and eddies in the bottom 100 m is highly significant. Detailed analyses of the curves suggest that the bottom 100 m is sometimes homogeneous, implying rapid mixing within it.

Fronts in continental shelf waters

Fronts occur where water masses of distinctly different properties meet (Bowman and Esaias, 1978). They are best viewed as two-dimensional planes that are often nearly vertical, and across which water properties change rapidly. Most fronts are characterised by waters of different salinities and temperatures converging towards them from both directions, and then sinking at the front. The sinking is caused by *cabbeling*: a mixture of two waters having different salinities and temperatures but equal densities has a greater density than the isolated components, and so sinks.

A number of different frontal systems are recognised (Figure 2.14) including upwelling, river plume, shelf break, and shallow-sea fronts, and high biological activity is often associated with them.

River plumes form just offshore from some rivers and are caused by a layer of light, less saline water spreading over more saline water. The plume itself is slightly elevated in relation to sea level. The effect is very similar to a salt wedge in an estuary (Figure 8.3), but here the surface water of the plume mixes downwards at the front, not upwards. *Upwelling fronts* occur where water that has upwelled moves offshore and meets less saline warmer offshore water and sinks (Figure 2.13).

Shelf break fronts occur at the edge of the continental shelf and are well documented on the east coast of North America (Mooers, Flagg, and Boicourt, in Bowman and Esaias (eds.), 1978). Their structure is most easily seen in winter. South of Rhode Island (east coast, USA), for example, sea surface temperature is about 4 °C, near the shelf edge it is 8 °C, and beyond the shelf 15 °C. The equivalent salinities are about 31‰, 34‰ and 36‰. The near-shore water is therefore much colder and less saline than the water beyond the shelf. In addition, there is a marked discontinuity, in otherwords an area of rapid salinity and temperature change, at the edge or break of the continental shelf. This discontinuity is called a shelf break front and develops above the point at which the continental shelf merges into the continental slope (the shelf break). The fronts are fairly stable in winter, but their position can be influenced by storms and upwelling, and they sometimes release bubbles of shelf water into slope water further offshore. The winter difference between the continental shelf and the slope water which leads to the establishment of shelf break fronts is thought to develop as follows. Shelf water is shallow. In autumn it cools rapidly and vertical mixing occurs throughout the water column. It also receives a lot of fresh water from estuaries. Hence the whole water mass is cold and has a relatively low salinity. In contrast, the offshore slope is so deep that water cooled at the surface by the cold winter air sinks immediately to much greater depths, where it is too deep to be mixed with the remaining warm surface water. In winter then, offshore slope water is temperature stratified

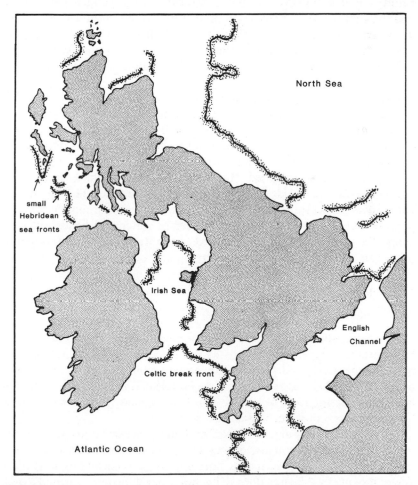

Figure 2.14 Shallow sea fronts around the British Isles detected by satellite infrared images (NOAA-4 satellite photographs taken during 1975, 1976, 1978) (after Simpson, in Cracknell (ed.), 1981.)

but continental shelf water is not, and a front develops between the stratified and unstratified water above the shelf break.

Shallow sea fronts are well documented in European continental shelf waters (Simpson and Hunter, 1974; Pingree, Bowmann and Esaias, and Simpson and Pingree, in Bowmann and Esaias (eds.), 1978). The presence or absence of stratification in these waters can be predicted from h/u^3, where h = water depth and u = amplitude of oscillations of tidal current velocity. Fronts are likely to develop where stratified and unstratified water abutt, and the contours of h/u^3 are good predictors of these fronts. Major fronts have been predicted and found in the Irish Sea area and the southern approaches to the English Channel (Figure 2.14), and are usually about 100 km long. Smaller fronts are also found in inshore waters near estuaries and headlands.

Biological productivity is often high in and near fronts. Materials brought by the converging currents often accumulate at the front and seabirds, fish, dolphins and whales are all known to aggregate there. Large commerical fisheries occur in and around shelf break fronts off the East coast of Canada and Newfoundland, and high concentrations of phytoplankton are also recorded. High densities of phytoplankton can also occur on the stratified side of shallow sea fronts in summer. This apparently contradicts the accepted dogma that nutrient depletion in summer reduces phytoplankton growth. However, transfer of water and hence nutrients across the front from the non-stratified to the stratified side is thought to occur very rapidly, thus stimulating phytoplankton growth on the stratified side.

Microstructure, advection, and diffusion

Divers have shown by releasing dyes that the ocean has a marked microstructure superimposed on the major currents. Differences in currents, temperature, and salinity, can occur within metres or even centimetres. Figure 2.15 shows a particular example of this effect on a somewhat larger scale of hundreds of metres. This heterogeneity has altered some of our conceptions of water movement in the sea.

The movement of water in the sea, both in the major ocean currents and in smaller-scale movements, occurs by advection and diffusion. Water moves from place to place either horizontally or vertically by advection. Water moving by diffusion does so by molecular diffusion, as would a blob of dye, or by eddy diffusion which can be compared with the blob of dye being stirred by a spatula. In the sea, eddy diffusion is caused by local

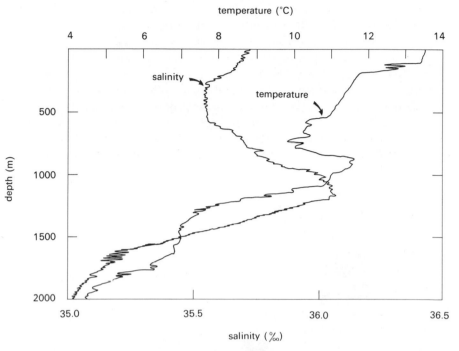

Figure 2.15 Microstructure of the ocean, as measured by a continuously-recording temperature-salinity-depth probe (modified from Herring and Clarke, 1971).

turbulence. Clearly eddy diffusion is a more rapid process than molecular diffusion, and advection more rapid than either.

Tides

Tides are most obvious at the shore where the sea level rises and falls regularly, usually twice a day. Vertically they can move less than a metre, as in the Mediterranean and around Jamaica, West Indies, or up to 15 m in the Bay of Fundy, Canada. Tides are caused by the movement of the moon around the earth, and to a lesser extent by the sun. The sun's tide-generating force is a little under half that of the moon when calculated from their relative masses and distances from the earth. The forces producing tides are very small, representing an acceleration of about 10^{-4} cm s^{-2}, but their effect is large when applied to the ocean as a whole. The earth and moon revolve around each other about the centre of their joint mass. This point is within the earth but not at its centre. The average distance between the centre of the earth and the centre of the moon is constant, and so the centrifugal force of the earth-moon system must be exactly balanced by the gravitational force of the system. The centrifugal force is constant all over the earth, but the gravitational force is greatest at the point on the earth's surface nearest the moon, and least at a point on the earth's surface furthest from the moon. At the nearest point, water will be pulled out into a bulge (the tide) towards the moon, because the moon's gravitational force is greater than the constant centrifugal force; while at the furthest point, water will be pulled in a bulge away from the moon because the constant centrifugal force

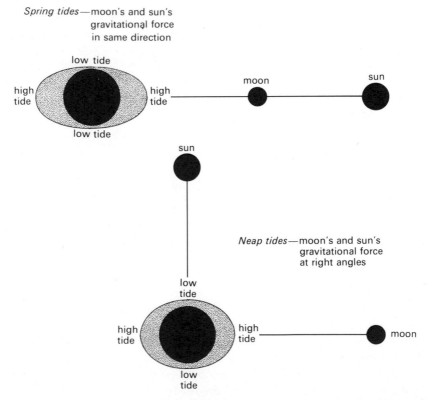

Figure 2.16 Tides are produced by the gravitational force of the moon, and to a lesser extent of the sun, acting on the oceans. When the gravitational forces of the moon and sun act in the same direction, the large amplitude spring tides are produced (upper diagram). When the forces act at right angles, the small amplitude neap tides are produced (lower diagram). Not to scale.

is greater than the moon's gravitational force (Figure 2.16). This gives rise to the two high tides per day which circle the earth as the earth revolves. The two low tides are about half-way in time between the two high tides. The period of the principal semi-diurnal tide is 12 h 25 min, which is equivalent to half the moon's apparent time of revolution around the earth. This means that successive high tides are separated by about 12.4 h. These statements are however very general since there is much geographical variation and there are additive effects caused by interactions between the sun and moon.

Spring tides have a large amplitude, and are caused by the moon's and sun's gravitational fields pulling in the same direction; this occurs when the moon is new or full. The smaller amplitude *neap tides*, in the moon's first and last quarter, are caused by the sun's gravitational force acting at right-angles to the moon's, and so the two forces have less effect (Figure 2.16). Spring and neap tides occur twice every lunar cycle of 28 days (the moon revolves around the earth once every 28 days). The influence of the sun's gravitational field is greatest at the time of the equinoxes, that is, when the sun is directly over the equator in March and September; during these periods the spring tides are particularly large.

Amphidromic points, seiches, meteorological conditions

The presence of estuaries, islands, and channels (e.g. the English Channel), and of semi-enclosed areas of sea (e.g. the Irish Sea, the Bay of Fundy) often have significant effects on the time and amplitude of local tides. For example, in the Irish Sea when the tidal current is flowing with its maximum speed northwards through St. George's Channel, the water surface is about 2 m higher on the Welsh coast than on the Irish coast, while the effect is reversed when the tide flows in the opposite direction.

Each area of sea has particular oscillatory characteristics. Tidal oscillations may revolve around a point of little tidal movement called an *amphidromic point*, or may oscillate to and fro to produce an oscillation called a *seiche*. Seiches are known to occur in the Bay of Biscay, and can also be produced by unusual meteorological conditions.

Co-tidal lines can be drawn radiating out from amphidromic points (Figure 2.17). These points having high tides at the same time, and usually show the time of high water in lunar hours (1 h to 2 min) after the moon's passage past the Greenwich meridian. *Co-range* lines can also be drawn circling amphidromic points. These join points having an equal tidal range.

The North Sea has a resonant period of about 40 hours depending on

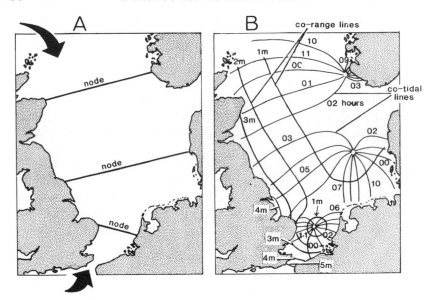

Figure 2.17 Tide generation in the North Sea. *A*, hypothetical three-nodal seiche. Large arrows show tidal waves entering from the north and south. The three lines represent the three hypothetical nodes. These form the three observed amphidromic points shown in *B*. This is caused by the action of the Coriolis force and the shape of the North Sea basin (after Harvey, 1976).

which axis you take. Tidal waves enter it from the North Atlantic Ocean, past the Orkney and Shetland Islands, and from the South through the English Channel. These set up a three-nodal system of seiches which are modified by the Coriolis force to form three amphidromic points: just west of Norway, west of Denmark and in the English Channel between England and Holland (Figure 2.17).

Meteorological conditions of high wind and abnormally high or low atmospheric pressure can also affect tides. Low atmospheric pressure, which is usually associated with poor weather, will allow sea level to rise. If it coincides with a spring high tide and an on-shore wind, the effects can occasionally be devastating, and are called a *storm surge*. Tides of 3 m above predicted levels occurred on 31 January 1953 on the east coast of England and caused widespread flooding and destruction. Like weather prediction, the prediction of tides is empirical and uses past long-term records to predict future tidal heights and times. Tides are measured by tide gauges which record the heights and times of tidal movements (Figure 2.18).

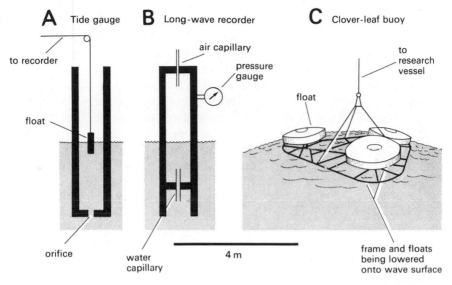

Figure 2.18 Tide gauge and longwave recorder, both operated near shore (modified from Hill, vol. 1, 1962). Clover-leaf buoy for recording the characteristics of waves over the deep ocean. The buoy is lowered onto the surface of the sea from a research vessel and the angle of the floats to the horizontal is electronically recorded on board (modified from Herring and Clarke, 1971).

Tsunami

Tsunami, misnamed *tidal waves,* are a succession of very fast low-amplitude long-wavelength waves which travel across oceans from volcanic explosions or earthquakes. In the open ocean, speeds of 750 km h^{-1}, wavelengths of 24 km, and wave amplitudes of less than a metre have been recorded. As they enter shallow water their speed falls because of friction with the bottom, and as a result their height may rise from a metre to 10 metres or more. They cause great loss of life when they flood low-lying islands such as Japan and many of the Pacific islands. It is interesting to note that the *tidal generating force* encircles the equator at 1700 km h^{-1}, that is, more than twice as fast as tsunami (Weyl, 1970). However, the oceans are ten times too shallow to have the same frequency as the tidal forces, and so no giant waves are produced.

Waves

Waves at the sea surface are usually produced by winds. Normally, a mixture of different waves moves in the wind's direction. The waves become more ordered into a swell when they move out of a windy area. In a moving wave in deep water, water particles move in a nearly circular orbit, rising to meet each wave crest as it passes them (Figure 2.19). The surface particles are crowded together at the wave crest and pulled apart at the trough. Waves move in the direction of the wind, but the individual particles of water do not, although there is a slow *mass transport* of surface water in the wind direction which causes the major ocean currents. The wind reinforces waves as follows: air is deflected up at the wave crest, causing relatively low pressure just in front of the crest. The water is moving upward as the wave moves to the right, and so is reinforced by the low pressure. Towards the next crest air is moving downwards again because the wave is moving to the right and is reinforced by the high pressure developing behind the wave. At about half a wavelength's depth, wave action is negligible (Figure 2.19).

Wave size and speed depend on wind speed, wind duration and the distance over which the wind blows. Theoretically a wave's height cannot

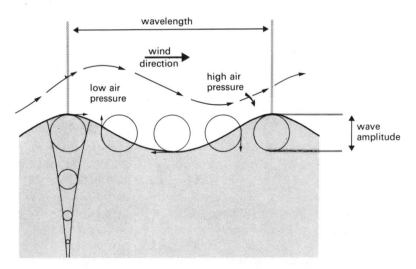

Figure 2.19 Wave characteristics and wind action at the sea surface. The orbits of individual water particles are shown by circles. Note that in the trough of the wave, the water particles are moving backwards while at the peak they are moving forwards (to the right). On the left the reduced influence of a wave away from the surface is shown by progressively smaller circles. The influence becomes negligible at a depth of about half a wavelength.

exceed about 0.14 of its wavelength. White caps and breaking waves in strong winds occur when this factor has been exceeded. As waves enter shallow water they slow down and shorten in length because of friction with the sea bed, and the wave crest eventually overtakes the body of the wave, becomes unstable, and breaks onto the beach (Figure 7.2). The energy of the wave's motion has been translated into energy of the water's movement up the beach. If waves enter shallow water at an angle other than 90° to the coast they become refracted or bent, and their direction of movement tends back toward 90°, that is, the waves tend to break parallel with the beach (Figure 7.5A). One result of this behaviour is that waves bend in towards headlands or promontories, and so concentrate their energy there.

Waves at the sea surface can be classified mathematically into two types depending on their wavelength, L, and on the water depth, h. *Long waves*, also known as *shallow water waves*, have a wavelength which is long compared to the water depth, and $h/L < 0.04$ ($< 4\%$). This means that a *long wave* is defined as one in which the wavelength is more than 25 times the water depth. Tides fall into this category. *Short waves*, also known as *deep water waves*, have a wavelength which is short compared to the water depth, and $h/L > 0.5$ ($> 50\%$). This means that a *short wave* is defined as one in which the wavelength is less than twice the water depth. Most waves at the sea surface obviously fall into this latter category. Intermediate conditions exist when $0.04 < h/L < 0.5$. Allen (1985) gives a good account of the mathematics.

The *wave velocity* (*phase velocity*) at the sea surface, C, and the resultant surface *water velocity* in the direction of the wave, U, are related to the wave's amplitude and wavelength and to water depth. However the relationships are different for *long* and *short waves*.

For *long waves*,

$$C = (gh)^{1/2}$$

where g = force due to gravity. The wave velocity therefore depends only on water depth, h, increasing as the square root of the water depth. The resultant surface water velocity is

$$U = a\left(\frac{g}{h}\right)^{1/2}$$

where a = half the wave amplitude. So the water velocity in the direction of the waves depends only on wave amplitude and water depth, increasing with increasing amplitude and with decreasing water depth.

For *long waves*, then, large amplitude waves in very shallow water produce fast water currents. Applying the equation for U to the North Sea, and taking $a = 2.0\,\mathrm{m}$ and $h = 30\,\mathrm{m}$, gives a peak tidal current of about $1.1\,\mathrm{m\,s^{-1}}$.

For *short waves*

$$C = \left(\frac{gL}{2\pi}\right)^{1/2}$$

The wave velocity is therefore independent of water depth and proportional only to the square root of the wavelength, L. The resultant surface water velocity is

$$U = \frac{a}{2}\left(\frac{\pi g}{2L}\right)^{1/2}$$

So the water velocity in the direction of the waves depends only on wave amplitude and wavelength, increasing with increasing amplitude and with decreasing wavelength. For *short waves*, then, large amplitude waves with short wavelengths produce fast water currents.

Ripples

Although we can now describe waves fairly accurately by scientific observation and measurement, theories on exactly how they are generated do not agree. However, very small waves or ripples are important. They determine the air/sea friction which in turn affects how the wind drives ocean currents and determines the rate of water evaporation and heat transfer to the earth's atmosphere.

Waves also occur below the sea surface. They usually have a longer wavelength and longer period but slower speed than typical surface waves. Internal waves have been detected at $300\,\mathrm{m}$.

Wave measurement

Surface waves are difficult to measure accurately. Near the shore they can be measured by apparatus such as the longwave recorder, while in the open sea the clover-leaf buoy is operated from a research vessel (Figure 2.18). Waves often move in groups across the sea surface, and the velocity of individual waves within a group is usually twice that of the group as a whole. Individual waves move through a group, reach the front of the group, and then die out. Groups of waves have recently been followed for very long distances across the Pacific by computer analysis. Wave periods,

Table 2.3 Characteristics of typical waves; the data are very approximate (modified from Hill, 1962, vol. 1.)

Wave type	Period (s)	Wavelength (m)	Maximum height (m)	Velocity of groups of waves (ms^{-1})
Long ocean ground swell	15	350	15	11
Ocean waves	7	80	4	5
Waves in anchorages	3	15	0.7	2.5
Ripples	< 0.5	< 0.05	< 0.005	< 0.05

the time from one peak to the next, vary from less than a second for ripples to 15 or 20 seconds for a long ocean swell (Table 2.3), while wave speeds vary from less than 10 cm s^{-1} to more than 25 m s^{-1}, and wavelengths from less than 1 cm to more than 400 m.

Generation and control of sea surface motion

All motion at the sea surface falls into a number of categories whose characteristics are generated and controlled by a range of forces

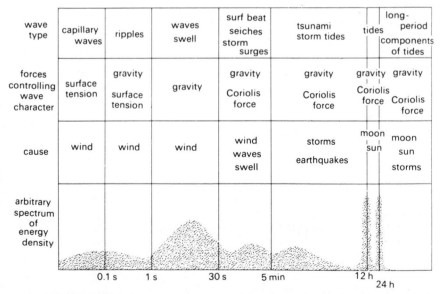

Figure 2.20 Categories of water movement at the ocean surface with their causes and controlling mechanisms. The period of the waves is plotted along the x-axis (modified from McLennan, 1965).

(Figure 2.20). Ripples and waves are generated by the wind and controlled by surface tension alone or in conjunction with gravity. Seiches, tsunami and tides are generated by wind, by earthquakes, and by the sun's and moon's attraction, and are controlled by gravity and the Coriolis force. The diurnal tides and ordinary wind-induced waves and swell contain the most energy.

CHAPTER THREE

PHYSICS AND CHEMISTRY OF THE OCEANS

Pressure

Pressure in a liquid increases rapidly with depth. Hence in the sea, a surface swimmer using a snorkel cannot breathe at depths greater than about 50 cm because the pressure difference between the surface and 50 cm depth is too great (c. 0.05 atm or about 4 cm Hg). The atmospheric pressure at sea level is called *one atmosphere* and is caused by the weight of air above. This weight is on average 1.033 kg cm^{-2} or 14.7 lb in^{-2}, and is equivalent to the weight of a column of mercury 760 mm high. Atmospheric pressure is also measured in bars.

One bar $= 1000$ millibars (mb) $= 1\,000\,000$ dynes cm$^{-2} = 10^5$ pascals (Pa) $= 10^5$ newtons m^{-2} (N m^{-2}).

The average pressure at sea level is 1013 mb.

One atmosphere $= 760$ mm Hg $= 1013$ mb $= 1.033$ kg cm^{-2}.

In water, pressure is directly related to water depth and increases one atmosphere per 10 m. At 10 m it is therefore $1 + 1 = 2$ atm, at 30 m it is $1 + 3 = 4$ atm, at 200 m it is $1 + 20 = 21$ atm, and so on, the 1 atm being atmospheric pressure at sea level.

SCUBA divers are very much affected by pressure. If they hold their breath at the surface and dive to a depth of 10 m, the ambient pressure will have increased from 1 to 2 atm, and the volume of air in their lungs will have decreased, therefore, by one half or 50%. This problem does not arise when a diver breathes compressed air from SCUBA (self-contained underwater breathing apparatus) because the demand valve that supplies air to the diver from the SCUBA air cylinders does so at the same pressure as the surrounding water. However the problems of nitrogen narcosis and oxygen poisoning, and of decompression sickness and air embolism are likely to be encountered if dives are not conducted to a strict schedule. These are best understood as follows.

Air pressure at sea level is about 1 atmosphere. The partial pressures (p.p.) of oxygen and nitrogen at sea level are therefore about 20% and 80%

of 1 atm, or 0.2 and 0.8 atm respectively. At 10 m depth the total pressure is 2 atm (see above), and so the partial pressure of oxygen is $2 \times 0.2 = 0.4$ atm and the partial pressure of nitrogen is $2 \times 0.8 = 1.6$ atm. At 50 m (c. 160 ft) the total pressure is $5 + 1 = 6$ atm, and the O_2 p.p. is 1.2 atm and the N_2 p.p. is 4.8 atm. Nitrogen at about this pressure, 4 to 5 atm, or 4 to 5 times the total pressure of air at sea level, induces a feeling of dizziness called nitrogen narcosis, and reasoning and decision-making are impaired. Experienced divers breathing air can work at 50 m with difficulty. At 90 m (c. 300 ft) the total pressure is $9 + 1 = 10$ atm, the O_2 p.p. is $10 \times 0.2 = 2$ atm and the N_2 p.p. is $10 \times 0.8 = 8$ atm. At this depth, divers breathing air experience very marked nitrogen narcosis and may suddenly lose consciousness due to oxygen poisoning caused by the high partial pressure of this gas. Needless to say, 90 m is well outside the normal limits of compressed-air diving. Oxygen and helium mixtures are now used for deep diving, on oil rigs in the North Sea, for example, and these mixtures to some extent ameliorate the dangers of breathing air at high pressures.

Decompression sickness and air embolisms are two well-known pheno-mena that are associated with incorrect diving procedures. Decompression sickness is caused by too rapid ascent from deep water, which results in the formation of bubbles of inert gas (mainly N_2) in the body tissues; it is painful, crippling, and sometimes fatal. Mild symptoms are aches and pains in the joints. Air embolism is caused by divers ascending to a shallower depth while holding their breath, thus producing air bubbles in the blood circulation (rather like suddenly removing the top from a bottle of soda water). Unconsciousness is often immediate and death frequently ensues. Modern treatment for decompression sickness and air embolism involves immediate administration of pure oxygen and recompression to 3 atmosph-eres in a pure oxygen environment.

Illumination

Sunlight is rapidly absorbed by sea water (Figure 3.1). Daylight falls to 1% of its surface value at about 100 m in clear oceanic waters, while in coastal waters the 1% level is reached at 10 to 30 m and in very turbid inshore waters or estuaries at less than 3 m.

These differences emphasise that inshore waters are cloudy compared to oceanic waters off the continental shelf. The cloudiness can be caused by concentrations of plankton, especially in semi-enclosed areas like the west-coast sea lochs of Scotland, or by suspended organic and inorganic material in estuaries. The exact measurement of attenuation of light under water

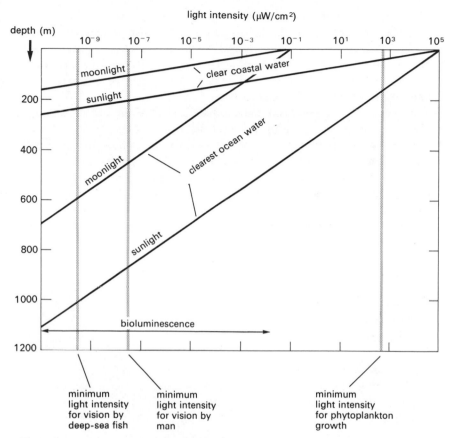

Figure 3.1 Attenuation of sunlight and moonlight with depth in clear coastal waters and in oceanic waters. Light intensity along the top axis is plotted on a $\log_{10}$ scale (modified from Hill, vol. 1, 1962).

requires complicated equipment, but a rough value can be obtained from the distance at which a white circular *Secchi disc*, 30 cm diameter, is last visible when lowered below the surface. In clear coastal waters some light is detectable at more than 200 m deep, and in the clearest oceanic water at 1000 m, but these light levels are below the resolution of the human eye.

Spectral composition of light in sea water

Most radiation in the ultraviolet and infrared range of wavelengths is rapidly attenuated even by distilled water, and this is much more marked for natural sea water (Figure 3.2). Coastal waters are most transparent in the yellow-green part of the light spectrum (wavelength 500 to 600 μm) while the clearest ocean waters in the tropics are most transparent in the blue region (400 to 500 μm). This accounts for the yellow-green appearance of underwater objects in coastal waters, and the blue appearance of underwater objects in the open ocean.

Photosynthetic pigments of marine plants, such as the benthic shallow-water seaweeds and floating dinoflagellates and diatoms in the phytoplankton, absorb maximally within the wavelength bands that penetrate furthest into the sea (400 to 750 μm) and so are well adapted to the marine environment (Figure 3.2). The rapid attenuation of light in sea water makes underwater exploration and photography by SCUBA divers and submer-

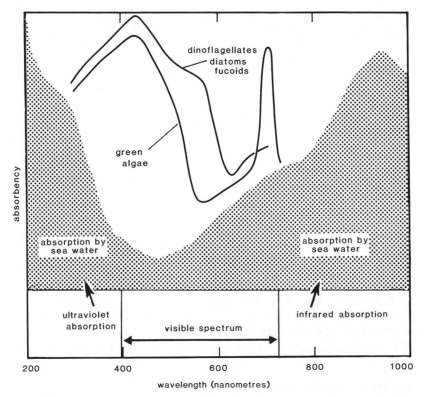

Figure 3.2 Rapid absorption of infrared and ultra-violet light by sea water, and the light absorption curves of the photosynthetic pigments of some marine algae (modified from Herring and Clarke, 1971, and Hill, vol. 1, 1962).

sibles difficult, and it is usually impossible to photograph objects without a photographic flash gun, or more than 10 m away even under optimum conditions. The scattering of light by suspended material and plankton often reduces the effective distance to 1 m in coastal waters.

Sound

Sound in the sea is produced by wave motion at the air/sea interface, by sediment transport, and by a wide variety of animals including man. Sound travels much more quickly and further in water than in air. At 20 °C its velocity in sea water is 1519 m s^{-1} compared with about 346 m s^{-1} in air (Sverdrup, Johnson and Fleming, 1942; Craig, 1973; Groves and Hunt, 1980). Since wavelength = velocity/frequency, sound of the same frequency will have a different wavelength in air and sea water. For example, within the frequency range of man's hearing (25–15 000 cycles per second), a sound having a frequency of 100 cps has a wavelength of 3.46 m in air and 15.19 m in water. The equivalent wavelengths for 1000 cps are 0.346 m and 1.519 m and for 10 000 cps are 0.0346 m and 0.1519 m.

Electromagnetic radiation as light or radio waves does not travel far in water and so sound has been used in the development of most remote sensing devices under water. These devices include echo sounders and side-scan sonars for recording the detailed topography of the sea bed, and pingers attached to equipment which allow the equipment to be relocated or to be released from the sea bed.

The velocity of sound in sea water is independent of wavelength except for very loud noises, but changes significantly with temperature, salinity and pressure. This means that sound velocity under water has to be calculated for each set of conditions, and tables are available for this purpose. In particular, sound velocity changes with depth. At the surface it is about 1500 m s^{-1}, it falls to about 1480 m s^{-1} at between 500 and 1000 m depth, and then increases to about 1525 m s^{-1} at about 4000 m. This means that a beam of sound transmitted at an angle to the horizontal will be refracted: in depths shallower than about 500 to 1000 m the beam will bend downwards and below these depths it will bend upwards.

The base of the thermocline is a unique sound transmission layer. The effect of the higher temperature above it and the lower temperature below it, together with the progressively increasing water pressure from the surface of the sea downwards, causes sound to be focused back towards the layer from above and below. Sound can be projected along this *deep sound channel* or *SoFar Channel* over very long distances. For example, depth charges exploded in the SoFar channel off Australia in the Pacific in 1960 were detected near Bermuda in the Atlantic, 19 000 km away.

Generally speaking, if there is no absorption or refraction, the intensity of sound varies inversely as the square of the distance from the source. However the picture is rather complicated in sea water. Absorption increases with increasing sound frequency and to a lesser extent with increasing viscosity of sea water. A good mathematical treatment is given by Sverdrup, Johnson and Fleming (1942).

Many animals produce sound under water. As coastal waters are approached in the tropics and subtropics, aggregations of different shrimp species produce sounds that *en masse* sound like the crackle of burning

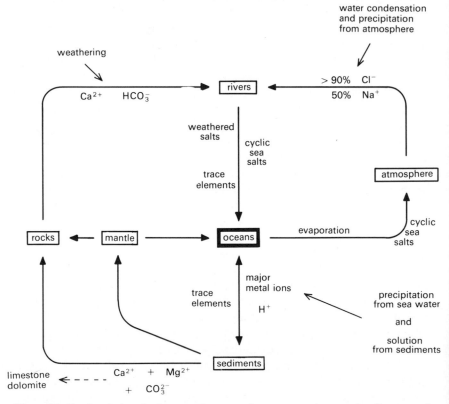

Figure 3.3 Geochemical cycles between the atmosphere, oceans, rivers, and sediments and rocks. Note the distinction between cyclic sea salts and salts liberated from rocks by weathering (see text). Many similar diagrams could be drawn for other elements and ions (modified from MacIntyre, 1970).

twigs and then further inshore like the sizzle of frying fat. Amongst fish, croakers and drum fish produce noises like banging on a hollow log in cycles of 4–7 blows. Other biological noises can sound like a cacophony of clicks, squeaks, honks, groans, barks and whistles, obviously produced by a very wide range of species. These include lobsters, shrimps, fish, porpoises, whales, seals and sealions (Groves and Hunt, 1980).

Inorganic ions in sea water

Many inorganic ions are present in solution in sea water. Overall and individual concentrations remain roughly constant, while many of them move to and from the land, sea and atmosphere in geochemical cycles (Figure 3.3). They are brought to the sea by rivers, and may be lost during evaporation from the sea surface (droplets, spray) and by precipitation in sediments. Sodium (Na^+) and chloride (Cl^-) are the commonest ions followed by the cations magnesium (Mg^{2+}), calcium (Ca^{2+}), potassium (K^+) and strontium (Sr^+), and the anions sulphate (SO_4^{2-}), bromide (Br^-), and bicarbonate (HCO_3^-). These ions represent 99.9% of the total inorganic salts of the sea ($35\,g\,l^{-1}$). We will consider the minor or rare ions in sea water later.

Sea water and river water, cyclic sea salts

Weyl (1970) has compared sea water and an average river water with relation to the cycling of inorganic ions. Some rivers contain many dissolved solids—the Colorado River contains 360 ppm (parts per million), while some contain few—the Columbia River contains 120 ppm. In general, however, the major ions in rivers are calcium and bicarbonate, and in sea water sodium and chloride (Table 3.1); so river water is not merely very dilute sea water. Now, spray droplets produced by waves and wind evaporate in air to form salt nuclei whose salt content is the same as sea water. The droplets range from about $10\,\mu m$ to $200\,\mu m$ diameter and contain between 10^{-11} and $10^{-6}\,g$ of salt (Woodcock, in Hill, vol. 1, 1962). When semi-dry, the salt particles are of the same order of size as the droplets. Winds carry them into the atmosphere, where water vapour condenses around them to form cloud droplets, and eventually rain. The rain falls either on the sea, thus returning the salt to the sea, or on the land, where some of the salt is carried in run-off to rivers and then back to the sea. These salts are called *cyclic* because they move in a cycle from the sea to the atmosphere, to the rivers, to the sea. The proportions of ions in cyclic salts

Table 3.1 Major inorganic ions in sea water and river water. Figures for river water are averaged from the world rivers. For the solutions to be electrically neutral, the concentration of cations must equal the concentration of anions in each liquid. Note that the relative percentages of ions in the cyclic sea salts in river water are about the same as those in sea water (after Livingston, 1963, and Weyl, 1970).

Sea water		Total $=$	River water rock weathering $+$	cyclic sea salts
Cations (%)				
Na^+	77	19	7	79
Mg^{2+}	18	24	25	17
Ca^{2+}	3	53	63	4
K^+	2	4	5	< 1
Total cations (m.eq kg^{-1})	605	1.42	1.18	0.24
% of sea water	100%	0.23%		
Anions (%)				
Cl^-	90	15	< 1	92
SO_4^{2-}	9	17	19	8
HCO_3^-	1	68	81	< 1
Total anions (m.eq kg^{-1})	605	1.42	1.18	0.24
% of sea water	100%	0.23%		

are the same as those in sea water. The fall-out of salt from the whole world atmosphere is about 2×10^9 tons per year.

Weathering of rocks contributes about 80% of the ions in river water, mainly magnesium, calcium, bicarbonate, and sulphate, while cyclic sea salts contribute the remaining 20%, mainly as sodium and chloride (Table 3.1, columns 3, 4). River water, therefore, contains cyclic sea salts from rain, and rock salts from rock weathering. The details of these processes are not fully understood.

Residence times of water and inorganic ions

The residence time of water and of salts in the ocean is important in relation to these cycles. The oceans contain about 1.4×10^{21} kg of sea water, and water flows into the oceans at about 3.14×10^{16} kg per year from the world's rivers. Dividing the former by the latter gives about 4.4×10^4 years, or 44 000 years. This is the time taken for river flow to replace the volume of the ocean completely and is called the *residence time* of water in the ocean.

Since rivers contain about 1.18 meq kg^{-1} of salt excluding recycled sea salt, and the seas about $605\,meq\,kg^{-1}$, the sea is $605 \div 1.18 \doteq 500 \times$ the ionic concentration of the average river. The residence time of total salt in the sea is therefore $500 \times 44\,000 = 2.2 \times 10^7$ years, or $22\,000\,000$ years.

The approximate geological age of the ocean is more than 10^9 years or a thousand million years. The residence times of water, total salt, and of individual ions, such as bicarbonate (0.11×10^6 years) and sodium (260×10^6 years) are much shorter than the geological age of the ocean. Salts, therefore, are not just being added to the sea and not removed; otherwise the residence times of the ions (time to replace them completely) would be about the same as the age of the ocean. The concentration of ions in sea water remains constant, so salts must be lost and replaced continuously, the ones with the shortest residence time being lost and replaced most quickly, and the whole being a dynamic equilibrium.

Ocean mixing times

Now let us turn to mixing times of water in the sea, that is, the time taken for a water mass once it has left the surface to be circulated to the ocean deep and then returned to the surface again. These times are measured by a radioactive method and are very approximate. Cosmic rays interact with atmospheric nitrogen to give radioactive carbon, ^{14}C. The ^{14}C enters the ocean, and once it has left the surface waters cannot be increased by addition from the atmosphere. The rate of decay of ^{14}C to ^{12}C (the stable isotope of carbon) is known; its concentration is halved every 5730 years (its half-life). By measuring the ratio of ^{14}C to ^{12}C in surface waters and the lower ratio of ^{14}C to ^{12}C in deep waters we can calculate how long the deep water has been out of contact with the sea's surface. For example, if the deep water contains exactly half the ^{14}C of surface water, it has been out of contact with the atmosphere for 5730 years. Calculated in this way, the mixing times of deep water are between 100 and 1000 years and so are much shorter than the residence times of the major ions in sea water. The ions added to sea water by the rivers are therefore very quickly mixed with those already present, which helps to explain the constancy of the ionic composition of the oceans.

Utilisation of inorganic ions by animals, plants and man

There are many other inorganic ions present in sea water but all at very low concentrations. Some of them, such as aluminium, iron and silicon are extremely abundant in the earth's crust and in rivers, but are very quickly removed in the more alkaline waters of the sea by sedimentation. Most of them have residence times shorter than the ocean's mixing time and so their concentration varies from place to place. Some are concentrated to a remarkable degree by marine animals and some are valuable enough for extraction. Vanadium is concentrated 10^5 times by tunicates, iron 10^5 times and chromium 10^4 times by algae, and zinc and copper 10^6 times in fish bones. The extraction of uranium from sea water is now at a pilot stage, and manganese nodules on the ocean floor are already being mined for their mineral content. Drinking water, chlorine, sodium, magnesium and bromine are commercially extracted from sea water at present, and many more elements and compounds will be extracted as the price of raw materials from other sources continues to rise.

Atmospheric gases in sea water. Factors affecting carbon dioxide and oxygen concentrations

All atmospheric gases are present in sea water, but with the exception of carbon dioxide at only 1 or 2% of their concentration in air (Table 3.2); this is even lower than their concentration in fresh water because of the salt in sea water. The value for carbon dioxide is about sixty times that in air, as it includes dissolved gas (CO_2), undissociated carbonic acid (H_2CO_3), dissociated bicarbonate (HCO_3^-) and carbonate (CO_3^{2-}). Gas concentrations are affected by the following factors. (a) Temperature and salinity influence the flow of gases across the air/sea interface; for example, oxygen and nitrogen solubilities fall when temperature and salinity rise. (b) Photosynthesis and respiration alter the levels of oxygen and carbon dioxide. (c) Water mixing changes the surface waters.

Surface waters are often saturated with O_2 and sometimes *supersaturated* (see below). Between 150 and 1000 m deep, an oxygen minimum layer sometimes forms which is about 100 to 200 m thick and sometimes under 2% saturated. It is most obvious near the equator. This layer contains many animals, and its low oxygen content may therefore be caused by animal and bacterial respiration. Below the oxygen minimum layer the oxygen level rises again. In some enclosed waters, such as deep lochs with sills, fjords, and areas such as the Black Sea which is separated from the Mediterranean by the shallow Bosporus, the deep bottom waters may be deoxygenated or even anoxic. In these conditions, run-off from rivers produces low salinity, low density, surface water which will not mix with the more dense, high salinity water below. Often animals cannot live in these deeper waters and anaerobic bacteria flourish.

Table 3.2 Gases in the atmosphere and oceans (after Hill, vol. I, 1962; Sverdrup, Johnson and Fleming, 1942).

	Atmosphere (ml/l air)	Oceans (ml/l sea water)	% of total gas on earth in atmosphere	in the oceans
Nitrogen	781	8.4–14.5	99.4	0.6
Oxygen	210	0–8.5	99.5	0.5
Argon	9.32	0.2–0.4	98.8	1.2
Carbon dioxide	0.3	34–56*	not applicable	

*CO_2, H_2CO_3, HCO_3^-, CO_3^{2-}

Note: Helium, krypton, neon and xenon are also present in sea water, but in minute quantities.

Sea water is almost saturated with nitrogen regardless of depth, and so fixation or production of nitrogen by microorganisms is probably unimportant.

Carbon dioxide is present in equilibrium with its various ions:

$$\xleftarrow{\text{lowered pH}}$$
$$\text{addition of acid}$$

$$\text{dissolved } CO_2 \rightleftharpoons H_2CO_3 \rightleftharpoons HCO_3^- \rightleftharpoons CO_3^{2-}$$

$$\text{addition of alkali}$$
$$\xrightarrow{\text{raised pH}}$$

At its normal pH of 8.0 to 8.3 sea water contains mainly bicarbonate (HCO_3^-). The weak acids carbonic acid (H_2CO_3) and boric acid (H_3BO_3) with their dissociation products make sea water a fairly good buffer. If strong alkali is added (NaOH), more carbonic acid and boric acid are dissociated, and the pH remains constant until all the carbonic and boric acid are used up. The reverse applies on the addition of strong acid. The extreme range of sea water pH is a about pH 7.6 to 8.3. The measurement of sea water pH is difficult because of temperature and salinity effects. Increased temperature or pressure alters the dissociation constants of carbonic acid and causes a fall in pH.

Gas exchange across the air/sea interface

Rates of exchange or *flux*, F (moles m^{-3} yr^{-1}), across the air/sea interface depend on the concentrations of a given gas in the atmosphere (g_a) and sea water (g_s) next to the interface, the gas's *coefficient of molecular diffusion* (D), and the thickness of the *stagnant boundary layer* (Z).

$$F = \frac{D(g_a - g_s)}{Z}$$

For most gases in sea water, D is 2 to $18 \times 10^{-2}\,m^2\,yr^{-1}$, and Z is usually less than $50\mu m$. The thickness of the *boundary layer* is reduced by wind; this produces a larger gradient of gas concentration $[(g_a - g_s)/Z]$ and thus more gas exchange on windy days.

Now suppose $Z \simeq 20\,\mu m = 2 \times 10^{-5}\,m$ and $D \simeq 10 \times 10^{-2}\,m^2\,yr^{-1}$. Then $D/Z \simeq 5 \times 10^2\,m\,yr^{-1} = 500\,m\,yr^{-1}$ or about 1 to 2 m per day. Broecker (1974) interprets this as the top 1 to 2 m of the sea exchanging its gas with the atmosphere every day, or the top 100 to 200 m every 10 days if the sea is well mixed.

The flux of O_2 across the air/sea interface is $c.$ 600 moles $m^{-2}\,yr^{-1}$, and of CO_2 $c.$ 17 moles $m^{-2}\,yr^{-1}$. Since 1 mole gas = 22.4 l at N.T.P. these figures are $c.$ $1.34 \times 10^4\,l\,m^{-3}\,yr^{-1}$ respectively. Surface sea water is always saturated with O_2 because the O_2 flux is high in relation to O_2 production by plants and O_2 use by animals and plants. Indeed it is often slightly *supersaturated*. This is probably caused by entrainment of air bubbles rather than by phytoplankton photosynthesis because similar supersaturations are found for Ar and N_2.

Carbonate cycle, oolites and whitings

Bicarbonate is a major ion in sea water, and also has the shortest residence time. It is part of the carbonate cycle which is complex and not fully understood (Weyl, 1970). The following is a summary of its more important parts (Figure 3.4). The atmosphere contains 0.03% carbon dioxide which strongly absorbs infrared radiation and so influences the heat balance of the earth. Carbon dioxide in the atmosphere moves to and from sea water across the air/water interface, depending on the saturation of the surface water. As we have already seen, in sea water it is in equilibrium with carbonic acid, bicarbonate, and carbonate.

Carbonate is precipitated in sea water as follows. (*a*) Calcium carbonate forms sand-grain-sized nodules (oolites) which produce tidal and shallow water sediments and bars around the Bahamas. (*b*) A fine calcium carbonate precipitate suspended in sea water (whitings) often forms over the Bahama Banks. Cold deep water, rich in carbon dioxide and calcium, upwells in this region. As it warms, carbon dioxide escapes to the air, the pH drops, and calcium carbonate precipitates. This probably explains the

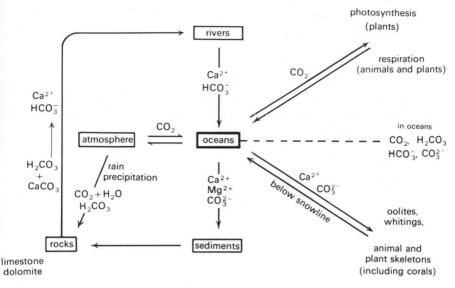

Figure 3.4 Carbonate cycle between the atmosphere, oceans, rivers, and sediments and rocks. The details are not understood as yet. Note that the cycle includes carbon dioxide (CO_2), carbonic acid (H_2CO_3), bicarbonate ions ($HCO_3{}^-$) and carbonate ions ($CO_3{}^{2-}$), as well as calcium (Ca^{2+}) and magnesium (Mg^{2+}) ions.

formation of oolites and whitings. (c) Carbonate, calcium and magnesium ions form limestone (Ca^{2+} and Mg^{2+}) in sediments. These latter then form consolidated rock and finally, under great pressure, marble. (d) Calcium carbonate is incorporated into animal and plant skeletons.

Carbonates are added to sea water in various forms: (a) carbon dioxide dissolves from the atmosphere; (b) calcium carbonate in dead plants and animals dissolves below the snow line (see below); (c) carbon dioxide is produced by animal and plant respiration; (d) bicarbonate with calcium enters in river water. Carbon dioxide is lost from the oceans to the atmosphere across the air/water interface and by photosynthesis. The concentration of carbon dioxide is never low enough to limit phytoplankton growth, in contrast with nitrate and phosphate, which have a marked limiting effect (see Chapter 4).

Carbon dioxide in the atmosphere is dissolved in rain as carbonic acid, which then dissolves limestone and dolomite to release calcium and bicarbonate ions into river water. Rivers then return calcium and bicarbonate ions to the ocean. Calcium carbonate is absent from sediments below about 4 km in the ocean (the *Snowline*). Whitish calcium carbonate shells of dead planktonic organisms sink through the water and settle on the bottom. Below about 4 km they dissolve and so sediments no longer look whitish. This solution of calcium carbonate is caused by a number of factors. The rise in pressure causes a fall in pH (see above); carbon dioxide production by animals reduces the pH and therefore the carbonate concentration; and the high pressure and low temperature increases the solubility of calcium carbonate.

Nutrients (phosphate, nitrate)

Plants in the sea need nitrogen and phosphorus as nitrate (NO_3^-) and phosphate (PO_4^{3-}) for growth (Chapter 4). These chemicals with other trace ions are called *nutrients*. Nitrogen in sea water is mainly present as nitrate, while microbial activity produces some nitrite (NO_2^-) and ammonia (NH_4^+) near the sea surface and near bottom sediments. In general, nitrate and phosphate levels are low in summer and high in winter. They are low and variable at the sea surface, increase with depth, reach a maximum at between 500 and 1500 m, and then fall again (Figure 3.5). The low surface levels in summer are caused by phytoplankton growth, and the

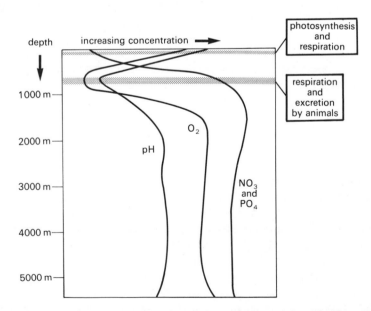

Figure 3.5 Oxygen, nitrate, phosphate, and pH changes with depth (modified from Harvey, 1957, Sverdrup, Johnson and Fleming, 1942, and Weyl, 1970).

Table 3.3 Concentration ranges of some nutrients and related substances in sea water. All concentrations in μg atoms per litre except Vitamin B_{12}. $1\,\mu g = 10^{-6}\,g$ (after Harvey, 1957; Hill, vol. 2, 1963; Sverdrup, Johnson and Fleming, 1942; Parsons, Takahashi and Hargrave, 1984).

Substance	Concentration
Particulate organic material (plankton and detritus)	10–500 μg C/l
Dissolved organic material	100–2000 μg C/l
Nitrate NO_3^-	0.01–50 μg N/l
Nitrite NO_2^-	0.01–5 μg N/l
Ammonium NH_4^+	0.01–5 μg N/l
Phosphate HPO_4^{2+}	0.01–4 μg P/l
CO_2 mainly as HCO_3^-	c. 2500 μg C/l
Vitamin B_{12}-like cobalamins	0.1–25 μg/l

maxima at intermediate depths are probably caused by animal excretion. The ratio of nitrate to phosphate is fairly constant, about 7:1 by weight, suggesting that the two ions are absorbed and released by living organisms in about the same proportions. Figure 3.5 shows diagrammatically the variation of oxygen, pH, nitrate and phosphate with depth. Near the surface oxygen and pH are high because of photosynthesis. Below the photosynthetic zone they fall, while nitrate and phosphate rise because of animal and microbial activity.

The concentration of inorganic ions (nitrate, nitrite, ammonia, and phosphate) varies widely because they are used by plants and produced by micro-organisms and animals (Table 3.3). Nutrient levels are best measured as soon as water has been collected by a reversing bottle. Oxygen is measured volumetrically, and nitrate and phosphate colorimetrically. Automatic equipment for use on board research vessels is now routinely used, which removes the tedium of these methods.

Particulate matter, vertical transfer, biolimited elements

Particulate matter sinking from surface waters towards the sea bed contains a wide variety of material ranging in size from c. 1 μm to c. 1 cm. It consists of (i) *organic matter*: secretions and faecal material from living organisms, and dead phyto- and zooplankton; (ii) *calcium carbonate*: skeletons of coccolithophorids, foraminiferans, pteropods and crustaceans; (iii) *silica*: skeletons of diatoms and radiolarians (Table 3.4). Its organic P:N:C ratio is relatively constant at 1:15:80 atoms and it contains much less carbon and

Table 3.4 Elemental composition of particulate matter compared with sea water (modified from Broecker, 1974). $-$ = low or negligible; (Particulate total + dissolved surface water) = dissolved deep water.

	P	N	C	Ca	Si
Particulate					
Organic	1	15	80	—	—
$CaCO_3$, SiO_2	—	—	40	40	50
Total	1	15	120	40	50
Dissolved					
Surface	—	—	680	3610	—
Deep water	1	15	680	3200	50

calcium than is dissolved in surface or in deep water. It is therefore an integral part of the carbon, nitrogen, phosphorus and silica cycles in the sea carrying these elements into deeper waters. This transfer is clear from the table, because the dissolved contents of deep water equal the dissolved contents of surface water plus total particulates.

Dissolved nitrogen (as NO_3^-), phosphorus (as PO_4^{3-}) and silicon (as SiO_2) are usually depleted in surface waters since phytoplankton use them in primary production; however, they are present in higher concentrations in deep waters since dead plankton sink and are dissolved by microbial and chemical activity. These nutrient-rich deep waters, when brought to the surface again by upwelling and wind action, produce the spring diatom bloom in temperate waters and the high productivity characteristic of upwelling areas (Figure 2.13, p. 28). Because they are biologically depleted in surface waters, nitrogen, phosphorus and silicon are called *biolimited elements*. Calcium, carbon, barium and radium are partly depleted by biological activity, and are hence called *bio-intermediate*. Elements that are not appreciably depleted are termed *bio-unlimited* (sodium, potassium, rubidium, caesium, magnesium, strontium, fluorine, chlorine, bromine).

Cycling of biological chemicals, the two-box model

It is sometimes convenient to consider the cycling of biologically important chemicals in terms of a two-box model in which the upper box is warm water above the thermocline and the lower box is cold water below the thermocline (Figure 3.6). We assume that dissolved chemicals enter only from the world's rivers and leave only by incorporation into particulate materials (animal and plant cells) and thence into sediments. This is a simplification, of course, since exchange also takes place across the air/sea

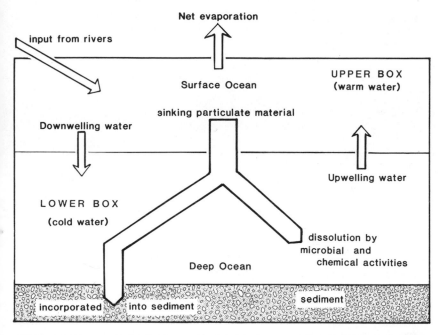

Figure 3.6 Diagram of the simple two-box model of ocean mixing (after Broecker, 1974).

and sea/sediment interface; however, the model helps to explain the main fluxes. The flux of any material is its rate of movement past a point, and is measured in units of material (g or moles) transported past the point/unit cross-sectional area of water/unit time. The major fluxes of biological chemicals occur by three mechanisms: in solution from the world's rivers, in solution in upwelling and downwelling waters, and as particulate material sedimenting through the water column and becoming incorporated into the sea bed.

Using this model it is possible to calculate for each ocean mixing cycle of 100 to 1600 years (i) the percentage of a chemical removed into particulate material (living cells) in the surface waters (Figure 3.6, upper box); (ii) the percentage carried into deeper waters as particulate material and not dissolved (lower box); and (iii) the percentage lost to the sediment as particulate material (Broeker, 1974). As an example almost 100% of the biolimited chemicals phosphorus and silicon are removed into living cells in surface waters, only 1% to 3% reach deeper water as particulates, and the same percentages then enter sediments (Table 3.5). The comparison with

Table 3.5 Vertical transfer of two biolimited and two bio-intermediate elements, and loss to sediments (modified from Broecker, 1974). Percentages are approximate.

		% removes as particulates (living cells)	% carried into deep water as particulates	% lost to sediments
Biolimited	P	93	1	1
	Si	97	3	3
Bio-intermediate	Ba	70	11	8
	Ca	1	16	0.2

the percentages for the bio-intermediate chemicals barium and calcium in the table are interesting.

Regeneration of nutrients from the sea bed

As organic matter sinks towards the sea bed and becomes incorporated into sediments, it is broken down by microbial action. This leads to the regeneration of nutrients such as ammonium, nitrate and sulphate from the organic matter.

Nutrient regeneration in coastal waters may be more important from the sediments than from the water column itself, and sediment regeneration is probably a major factor influencing primary production. Nutrient regeneration by zooplankton as urine and faecal material may be more important during zooplankton blooms and patches, but benthic regeneration is a more continuous process. Considerable regeneration occurs in the surface aerobic layers of sediments. However, the highest concentrations of ammonia are found in the subsurface anaerobic layers, and upward diffusion will occur from these. Bacteria are obviously important in this process but meiofauna such as ciliates may also play a major role. The contributions of microbial, meiofaunal and macrofaunal activity to nutrient regeneration in sediments and to nutrient fluxes across the sediment-water interface into the water column are not established as yet. There are, however, clearly defined maxima of various chemicals in sediments that by and large accord with microbial activity and Eh (Figure 3.7) and these chemical and microbiological gradients are almost certainly of great importance during early diagenesis in relation to redox potential profiles (Figure 3.10) and to diagenetic chemical zones and physiological groups of bacteria (Table 6.1, p. 114). It is difficult to generalise, however, because there is so much variability. The vertical scale over which the gradients occur can range from 0–1 cm in sediments on

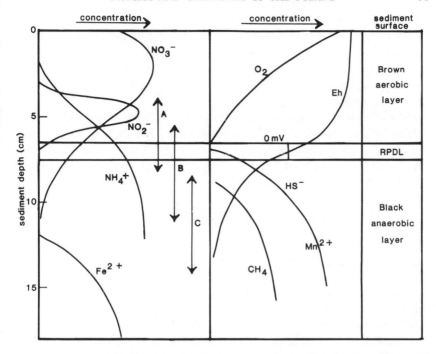

Figure 3.7 Concentration profiles of chemicals in sediments. RPDL = redox potential discontinuity Payer. Microbiological activity: *A*, nitrate reduction; *B*, sulphate reduction; *C*, carbonate reduction (methanogenesis) (after Anderson and Meadows, 1979; Krauskopf, 1979; Berner, 1980; Lewinton, 1982; Meadows and Tait, 1985).

muddy intertidal shows, to 0–50 cm in sediments on the outer continental shelf on slope. The exact position and relationships of the curves can also vary considerably, and the redox potential discontinuity layer does not even exist in abyssal plain sediments when the redox potential remains positive.

Humic and fulvic acids and kerogen in sediments

Humic acids are substances extractable from sediments by NaOH, and precipitable by acid at pH 2. *Kerogen* or *humin* is the stable residue after alkaline extraction and *fulvic acid* is the soluble material remaining in solution after acid precipitation. All three names are generic terms that cover a complex mixture of organic substances, produced by microbial degradation, chemical polymerisation, and oxidation (Poutanen, 1985).

They enter sediments either by sedimentation through the water column or by the activities of microorganisms and chemical changes in the sediments themselves. Their concentrations in the top 15 cm of sediments range from 2 to 100 mg g^{-1} dry sediments depending on hydrodynamic conditions in the overlying water, the source of the material, and the nature of the sedimentary environment.

Humic acids dominate when the organic matter in the sediment is of marine origin (offshore) and fulvic acids when the organic matter is of terrestrial origin (nearshore continental shelf). Both humic and fulvic acids are probably derived from unsubstituted lipids in plant tissue (land plants, macroalgae, phytoplankton).

Nitrogen and phosphorus cycle, microorganisms

Nitrogen and phosphorus in their various forms pass between different organisms in the sea. These movements can be drawn as cycles. The cycles are not quantitative or accurate, but they help to clarify the transfer of nitrogen and phosphorus from one source to another and to illustrate the importance of microbial activity in the sea.

(i) *The phosphorus cycle* (Figure 3.8). Zooplankton ingest organic and inorganic phosphorus in phytoplankton, and then excrete it in soluble form. Phytoplankton and microorganisms take in and excrete soluble inorganic and organic phosphorus. Organic phosphorus is hydrolysed to

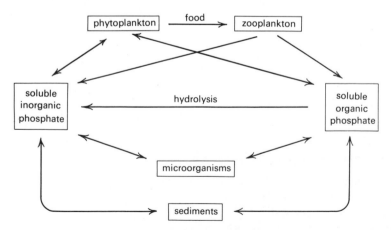

Figure 3.8 The phosphorus cycle in the sea. Zooplankton excrete inorganic and organic phosphate, and phytoplankton both excrete and assimilate inorganic and organic phosphate. The exchange of phosphate with sediments is not fully understood at present.

inorganic phosphorus by the alkaline pH of sea water and by phosphatases
on the surface of some phytoplankton. There is also interchange with the
sediment through the feeding and excretion of benthic microorganisms and
invertebrates. Phosphorus may be cycled about five times as quickly as
nitrogen.

(ii) *The nitrogen cycle* (Figure 3.9). Many parts are analogous to the
phosphorus cycle. Zooplankton ingest organic and inorganic nitrogen in
phytoplankton, and excrete it in soluble form as ammonium or organic
nitrogen. Phytoplankton and microorganisms take in and excrete soluble
inorganic and organic nitrogen. There is also interchange with sediments
through the feeding and excretion of benthic microorganisms and invertebrates. Phytoplankton can take up ammonium and nitrite ions, as well
as nitrate, and may also excrete nitrite during nitrate uptake. Blue-
green algae and some bacteria fix molecular nitrogen, especially in the
tropics. In temperate waters nitrate may be completely depleted, while
phosphate is still present at low levels. Bacteria play a major part in the
nitrogen cycle. They can fix molecular nitrogen, and also convert soluble
organic nitrogen, which has been produced by living phyto- and zooplankton and released from dead organisms, to ammonia (deamination of
amino acids). They oxidise ammonium to nitrite and nitrite to nitrate
(nitrification). These bacteria are mainly found in and just above bottom
sediments, and on plankton.

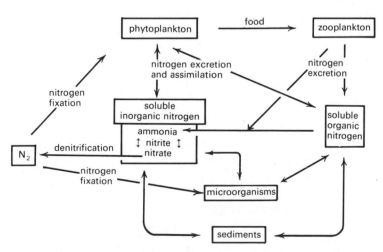

Figure 3.9 The nitrogen cycle in the sea. The exchange of inorganic and organic nitrogen with
sediments is not fully understood.

$$\xrightarrow{\hspace{2cm}} \text{nitrification} \xrightarrow{\hspace{2cm}}$$
$$\text{and oxidation}$$
$$NH_3 \longrightarrow NO_2^- \longrightarrow NO_3^-$$
$$\textit{Nitrosomonas} \qquad \textit{Nitrobacter}$$

The nitrite maximum that sometimes develops at the bottom of the photosynthetic zone may be caused partly by oxidation of ammonium to nitrite by bacteria and partly by the reduction of nitrate to nitrite by phytoplankton. Bacteria can also reduce nitrate to nitrite and nitrite to molecular nitrogen (denitrification).

$$\xrightarrow{\hspace{2cm}} \text{dentrification} \xrightarrow{\hspace{2cm}}$$
$$\text{and reduction}$$
$$NO_3^- \longrightarrow NO_2^- \longrightarrow N_2$$

The last step is probably unimportant in the sea, although bacteria which reduce nitrate to nitrite are common in sea water and in sediments.

Redox potential and pH of sea water and sediments

Many chemical, geological, and biological processes in sea water and sediments are related to the oxidation–reduction potential, or redox potential (Eh), and to the acidity (pH) of the environment (ZoBell, 1946; Baas Becking, Kaplan and Moore, 1960; Krauskopf, 1979). Oxidation–reduction potentials can be regarded as quantitative measures of the energy of oxidation or electron-escaping tendency of reversible oxidation–reduction reactions in the environment (ZoBell, 1946).

Eh is measured in millivolts (mV) with respect to a reference hydrogen electrode which is taken to have an Eh of 0 mV. Eh is positive in aerobic environments and negative in anaerobic environments. When the Eh is zero or negative the environment contains no free oxygen. A calomel ($Hg/HgCl_2$) reference electrode is normally used in place of the hydrogen reference electrode, whereupon a correction factor of c. 250 mV is added to the readings. Eh and pH electrodes need frequent calibration against standard solutions having a known Eh or known pH (buffer).

Relationships between Eh and pH in natural systems are conventionally presented on Eh–pH diagrams. Eh (y axis) is plotted as millivolts, usually ranging from -600 mV through zero to $+1000$ mV, and pH (x axis) is plotted in the normal pH units ($-\log_{10}[H^+]$), usually ranging from 2 to 12.

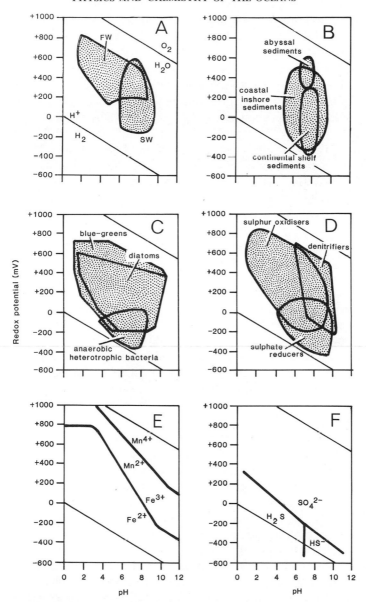

Figure 3.10 Eh–pH diagrams for *A*, fresh water and sea water; *B*, marine sediments; *C*, *D*, various marine phytoplankton and microorganisms found in sea water and sediments; *E*, *F*, stability fields for manganese, iron and sulphur ions (after Baas, Becking, Kaplan and Moore, 1960; Krauskopf, 1979).

Figure 3.10 shows Eh–pH relationships for a number of natural systems. Sea water is generally more basic and may have a considerably lower Eh than fresh water. Continental shelf and abyssal sediments have a slightly more restricted pH range than coastal inshore sediments but a slightly greater Eh range. Abyssal sediments have consistently higher Eh's than continental shelf sediments, presumably because their organic content and microbial activity is much lower. There are also obvious and predictable differences between photosynthetic algae and bacteria (diatoms, blue greens) and anaerobic heterotrophic bacteria (Figure 3.10C). The interesting Eh and pH limits of sulphate-oxidising, denitrifying and sulphate-reducing bacteria (Figure 3.10D) show that under certain conditions of Eh and pH, all three types may be active in sediments at the same time. These conditions are approximately $+100\,mV > Eh > -200\,mV$ and $7 < pH < 9$.

The vertical profiles of Eh and pH in sediments determine which chemical species are stable at particular depths (Figure 3.10E, F), and this is very important during early diagenesis in sediments (p. 116). For example, at pH 7.5 to 8.0, Mn^{4+} is stable above $+400\,mV$ and Mn^{2+} below $+400\,mV$, Fe^{3+} is stable above $+100\,mV$ and Fe^{2+} below $+100\,mV$, and SO_4^{2-} above $-220\,mV$ and H_2S on HS^- below $-220\,mV$. There are many such redox relationships between different oxidation states of elements which are exactly defined chemically, and which are of major importance in early diagenesis in the upper 1 to 2 m of most sediments.

Dissolved organic substances

Dissolved organic substances from living and dead plants, microorganisms and animals are present in the sea, although not a great deal is known about them. Carbohydrates, polysaccharides, fatty acids, peptides, amino acids, lipids and vitamins have all been identified. Vitamin B_{12}, for example, is very important for phytoplankton growth, and the huge phytoplankton population explosions called *red tides* kill many fish because of the release of poisonous external metabolites.

Some organic compounds are transferred between different levels of the marine food chain and these may be recycled. For example, in the Antarctic the phytoplankton *Phaeocystis* contains large amounts of acrylic acid. This chemical is strongly anti-bacterial. *Phaeocystis* is eaten by krill (*Euphausia superba*) which is eaten by penguins. Acrylic acid is transferred to the penguin and produces bacteriological sterility in the anterior part of its gut.

Yellow substances in sea water, *Gelbstoff*

Many natural waters contain *yellow substances* in solution, called *Gelbstoff*. These substances absorb energy in the ultraviolet and parts of the visible light spectrum from 350 to 500 nm. They have been isolated from algae, and are abundant in saltmarshes, mangrove swamps and estuaries. *Gelbstoff* is released into sea water by macroalgae and phytoplankton, and this occurs more quickly in the light than in the dark. The light-induced release is associated with a decrease in chloroplast pigment in the plant cells which suggests that *Gelbstoff* may be part of the pigment–protein complex in the living plant cell.

Gelbstoff is present at high concentrations in oceanic and coastal waters where rivers produce terrigenous discharge and lowered salinity. The discharge from the river Amazon, for instance, can be followed many km out to sea at progressively lower concentrations.

Yellow substances are probably removed from the water column by photo-oxidation, by adsorption on to particles that sink, and by microbial oxidation.

CHAPTER FOUR

THE PELAGIC ENVIRONMENT

This chapter describes the animals and plants that live in the pelagic environment. It begins by describing the phyto- and zooplankton, primary production and pelagic food webs. We then consider environmental seasonal and geographical effects on the abundance and productivity of the plankton, and follow this by an account of plankton patchiness and indicator species. The last part of the chapter covers neuston, pleuston and nekton, bathy- and abyssopelagic animals, and whales and sea birds.

Phytoplankton, zeoplankton, size, and sampling

Phytoplankton are the most abundant and widely distributed form of plant life on our planet and their annual net carbon production, 10^{10} tons, is at least 50% of global plant production (Smayda, 1970). They are microscopic floating plants that contain chlorophyll and hence obtain energy for growth by photosynthesis. They range in size from about 2 to 200 μm, and are diverse and sometimes bizarre in shape (Figure 4.1). Zooplankton are small herbivorous or carnivorous animals that feed on the phytoplankton or on other zooplankton. They range in size from about 20 to 5000 μm. Many of them, like the phytoplankton, are transparent and very beautiful. Some are amazingly abundant—copepods often make up 70% to 90% of zooplankton catches. The phyto- and zooplankton together represent the base of almost all food chains in the sea—either directly by being eaten, or indirectly by their breakdown products entering sediments and being eaten there.

Broadly speaking, both phyto- and zooplankton are limited to the upper sunlit zones of the sea. Phytoplankton live near the surface where there is enough light for photosynthesis (< 100 m approx.), although their resting stages are found deeper in the water column during winter. Zooplankton can live at much greater depths (see vertical migration, Figure 4.9). The skeletons of some phytoplankton form thick deep-sea sediments (e.g.

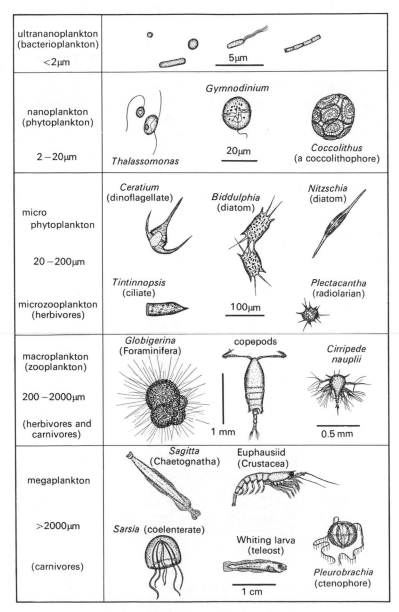

Figure 4.1 Phytoplankton and zooplankton.

globigerina ooze, diatom ooze, Figure 6.3) having sunk from the surface waters.

The smallest plankton are bacteria or related prokaryotes called the bacterio-plankton or ultra-nanoplankton ($< 2\,\mu$m diameter). Little is known of their biology. Small flagellates and coccolithophores constitute the nanoplankton (2–20 μm), all of which are also phytoplankton. The microplankton (20–200 μm) includes many phytoplankton, such as the dinoflagellates and diatoms, and some small herbivorous zooplankton, the ciliates and radiolarians. Larger plankton are all animals. Many invertebrate larvae (nauplii, trochophores), copepods, and larger Foraminifera make up the macroplankton (200–2000 μm), while the big carnivorous species such as *Sagitta, Pleurobrachia*, the euphausiids (krill) and many large invertebrate larvae (echinopluteus) make up the megaplankton ($> 2000\,\mu$m, i.e. > 2 mm). There are, of course, very many species and the size classification is only a rough one.

The distribution and abundance of plankton is measured by towing fine nets from research vessels (Figure 4.2). The Gulf III high speed plankton sampler is a metal funnel incorporating a flow meter and plankton net, and has a narrow mouth to prevent too fast a current from entering the net. The Hardy plankton recorder continuously catches plankton on a moving strip of gauze which is then stored in formalin. It can be towed for many

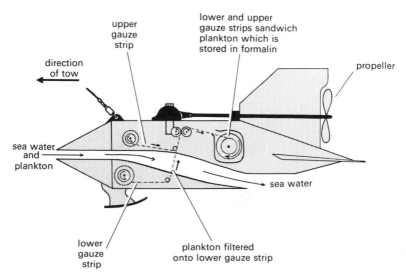

Figure 4.2 The Hardy continuous plankton recorder. The recorder is about 1 m long.

hundreds of kilometres by commercial vessels, and has given a unique picture of the North Atlantic plankton over many years. Other plankton nets such as the Indian Ocean Standard Net and the N.F. 70V are hauled through a given depth from a stationary research vessel and then closed.

Primary production, standing crop, measurement

Primary production is the first step in marine food chains or webs (Figure 4.3). It is the rate of production of new plant material by photosynthesis, usually measured as g carbon fixed per m² of sea surface per

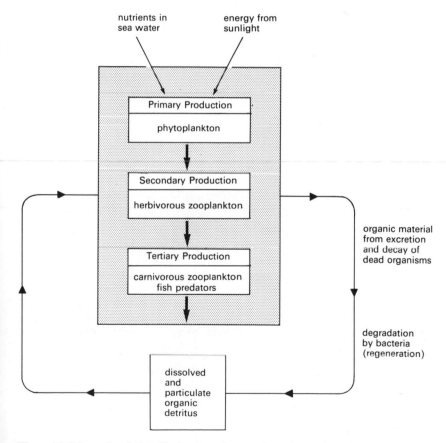

Figure 4.3 Schematic relationship between the recycling of organic carbon and primary, secondary and tertiary producers.

year, or per day. Since primary production is essentially the result of photosynthesis, it needs nutrients (nitrate and phosphate) and sunlight. On a global scale, marine primary production is limited by nitrate and phosphate. As a result, nitrate in the surface waters may be recycled between animals, heterotrophic microorganisms and phytoplankton 5 to 50 times per year, and phosphate 3 to 20 times (Russell-Hunter, 1970). Primary production by phytoplankton is estimated at about 25 to $35\,g\,C\,m^{-2}\,yr^{-1}$ by the ^{14}C method (see below) and about 100 to $200\,g\,C\,m^{-2}\,yr^{-1}$ by the O_2 method. In contrast, plants on agricultural land produce about $360\,g\,C\,m^{-2}\,yr^{-1}$ and the California seaweed kelp forests an astonishing $12\,000\,g\,C\,m^{-2}\,yr^{-1}$.

Phytoplankton *primary production* should not be confused with the phytoplankton *standing crop*. The standing crop is the amount of plant material present at any given instant in time $(g\,C\,m^{-2})$. It can increase even with a constant primary production if its losses are suddenly reduced. Grazing by zooplankton, for example, might decrease, or the quantity of phytoplankton being carried below the critical depth might be reduced. Both of these would increase the standing crop. The phytoplankton standing crop is measured as follows.

(i) Direct microscopic counts can be made on centrifuged or sedimented sea water.
(ii) Chlorophyll extracted from sea water by acetone can be measured colorimetrically. However a calibration curve of colour intensity of extract against number of plant cells has to be made beforehand.
(iii) The carbohydrate content of material filtered from sea water can be estimated. Zooplankton contains little carbohydrate, and phytoplankton a great deal. If feeding rates of zooplankton are known, zooplankton counts give an approximate estimate of loss of plant material to this source.

Primary production is usually measured by estimating oxygen release or carbon dioxide uptake as ^{14}C during photosynthesis. Photosynthesis proceeds as follows:

$$CO_2 + H_2O \longrightarrow (H_2CO)_n + O_2$$
$$\text{(uptake)} \qquad\qquad \text{(release)}$$

But phytoplankton also use oxygen and release carbon dioxide during respiration:

$$O_2 + (H_2CO)_n \longrightarrow CO_2 + H_2O$$
$$\text{(uptake)} \qquad\qquad \text{(release)}$$

Hence to measure photosynthesis, the oxygen taken up during respiration must be measured. This is done by measuring oxygen release (or carbon dioxide uptake) in the light and dark. Pairs of bottles, one of which is painted black, are filled with sea water that has been directly taken from the sea and hence contains phytoplankton, and are resuspended in the sea. Phytoplankton photosynthesis and respiration occur in the light bottle, but only respiration occurs in the dark bottle. The difference in oxygen released or carbon dioxide taken up (as $^{14}CO_2$) between the two bottles is then an estimate of net photosynthesis. The bottles can be suspended in the sea under ambient conditions, or incubated on board ship or on land under controlled light and temperature. Both methods have their advantages. There are also differences between the estimation of photosynthesis by oxygen release and $^{14}CO_2$ uptake. In the former, oxygen is estimated by the Winkler titration. In the latter, $^{14}CO_2$ is injected into the bottles at the beginning of the incubation, the phytoplankton filtered off at the end, and its ^{14}C content measured. Incubation times vary, but are usually 3 to 12 hours. The ^{14}C technique gives lower estimates of photosynthesis than the oxygen technique. There are problems in both methods in obtaining representative samples from patchy phytoplankton distributions, and in avoiding cell damage.

The oxygen method requires estimates of the *photosynthetic quotient* and the *respiratory quotient* (Levinton, 1982). The respiratory quotient (RQ) is the ratio (CO_2 molecules liberated)/(O_2 molecules assimilated) during respiration. The photosynthetic quotient (PQ) is the ratio (O_2 molecules liberated)/(CO_2 molecules assimilated) during photosynthesis and primary production of organic compounds. PQ = 1 for the formation of hexose sugars, 1.05 for the formation of protein from ammonia, 1.4 for the formation of lipid, and 1.6 for the formation of protein from nitrate.

Biological respiration in the sea

Overall estimates of biological respiration by living organisms in the sea are sometimes based on the presence of 10 moles of C to every 2 of N in organic material. 14 moles of O_2 are needed to respire this C and N to 10 moles of CO_2 and 2 moles of NO_3^-. Knowing the amount of organic detrital material present, it is therefore possible to predict changes in O_2, HCO_3^- and NO_3^- in sea water, and so to estimate how much of the observed changes are caused by respiration. For example, deep oceanic water has a lower O_2 level than predicted from equilibrium with the atmosphere at its *in situ* temperature, and this agrees well with the CO_2 increase produced by respiration of organic matter in the deep sea. The 14:2 or 7:1 ratio of $NO_3^-:O_2$ is also used as an NO marker of mass water movement in the oceans (page 27).

Food chains and webs

Phytoplankton synthesise organic substances using energy from the sun obtained by photosynthesis, and need nutrients such as nitrate and phosphate, inorganic ions, and carbon dioxide for growth. The plant growth, or primary production of new organic material (proteins, fats, carbohydrates), is the first link in all *food chains* in the sea (Figure 4.3). Herbivorous zooplankton eat phytoplankton, converting plants to animal tissue (secondary production), and they in turn are eaten by carnivorous zooplankton and fish predators (tertiary production). These are successive trophic levels in the *food chain* or *food web*. At each level, organic material is lost by excretion and by the natural mortality of those animals not eaten by the next stage. Bacteria are a vital link in breaking down organic materials for plant use again (regeneration). The efficiency of each step is very roughly 10%; in other words for 100 g food eaten, 10 g forms new body tissue. These processes can be regarded as an organic carbon cycle in the sea (Figure 4.3), in which carbon is cycled between the food chains and an organic detritus sink acted upon by bacterial decomposition processes. It is, of course, artificial to separate this cycle from the nitrate, phosphate, and CO_2 cycles (chapter 3), as all are intimately linked.

Pelagic food webs are complex and not fully understood. They involve many trophic levels—primary, secondary, tertiary, and so on—and are likely to be affected by many environmental variables such as light, temperature, salinity, and season. In the North Sea, adult herring feed on arrow worms, sand eels, the copepod *Calanus*, the amphipod *Themisto*, and the tunicate *Oikopleura*. These in turn feed on barnacle larvae, mollusc larvae, and small copepods which then feed on diatoms and flagellates. These interlocking food chains can only be described in very broad simplified terms at present (Figure 4.4). In addition, many species switch diets in the chain, while others feed at different trophic levels at the same time. Cod switch from eating crustaceans to herring when they reach a length of 50 cm, and some copepods eat detritus, phytoplankton, and small zooplankton.

In spite of these difficulties, the food chain concept (Figure 4.4) is a useful abstraction for summarising relationships. It is often possible to calculate the *ecological efficiency* of each level in the chain as (amount of energy

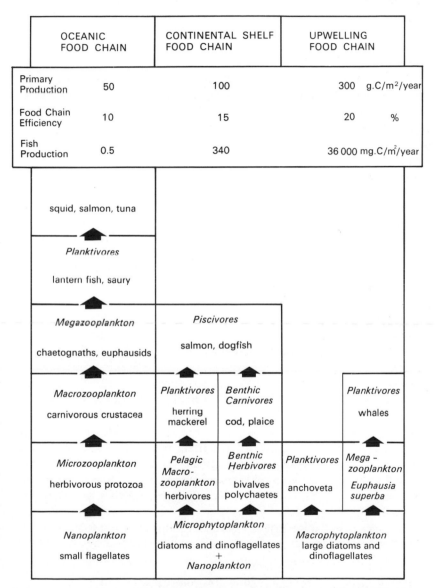

	OCEANIC FOOD CHAIN	CONTINENTAL SHELF FOOD CHAIN	UPWELLING FOOD CHAIN	
Primary Production	50	100	300	g.C/m²/year
Food Chain Efficiency	10	15	20	%
Fish Production	0.5	340	36 000	mg.C/m²/year

squid, salmon, tuna

▲

Planktivores

lantern fish, saury

▲

Megazooplankton   *Piscivores*

chaetognaths, euphausids   salmon, dogfish

▲   ▲   ▲

Macrozooplankton   *Planktivores* | *Benthic Carnivores*   *Planktivores*

carnivorous crustacea   herring mackerel | cod, plaice   whales

▲   ▲   ▲   ▲

Microzooplankton   *Pelagic Macro-zooplankton* | *Benthic Herbivores*   *Planktivores* | *Mega-zooplankton*

herbivorous protozoa   herbivores | bivalves polychaetes   anchoveta | *Euphausia superba*

▲   ▲   ▲   ▲   ▲

Nanoplankton   *Microphytoplankton*   *Macrophytoplankton*

small flagellates   diatoms and dinoflagellates + *Nanoplankton*   large diatoms and dinoflagellates

Figure 4.4 Ocean chains, continental food chains, and upwelling food chains (modified from Parsons, Takahashi and Hargrave, 1984; Ryther, 1969).

extracted from a trophic level) $\times 100/$(amount of energy supplied to that trophic level). Ecological efficiencies range from about 8–13% but this may be low because much of the remaining 90% is recycled to detritus and bacteria (Figure 4.3). If these ecological efficiencies are known, quantitative estimates of production at successive trophic levels can be calculated from $P = BE^n$, where B = primary production, E = ecological efficiency, and n = number of links between the first trophic level (primary production) and the trophic level corresponding to P (Parsons, in Cushing and Walshe (eds.), 1976). For example, supposing primary production is $5\,g\,C\,m^{-2}\,d^{-1}$ and $E = 20\%$, then production P at the second, third and fourth trophic levels ($n = 1, 2, 3$) will be 1.0, 0.2 and $0.04\,g\,C\,m^{-2}\,d^{-1}$ (Harden Jones, in Barnes and Mann (eds.), 1980). If the area of production and duration of production are known, annual production can then be calculated.

Ryther (1969) has used this approach to describe three major food chain communities associated with oceanic, continental shelf, and upwelling areas (Figure 4.4). The oceanic community has the lowest primary productivity, the largest number of trophic levels (nanoplankton ... piscivores), the lowest efficiency, and the lowest fish production. The upwelling community has the highest productivity, fewest trophic levels (macrophytoplankton ... planktivores), and highest efficiency and fish production. The continental shelf community is intermediate. Ryther's classification agrees with the very rich anchoveta fishery off Peru, with the good continental shelf fisheries, and with the poor oceanic fisheries.

The differences in food chain length may partly be explained as follows. On the continental shelf and in areas of upwelling, large chain-forming diatoms grow in the nutrient rich and often turbulent waters which are then eaten by large herbivores. This effect shortens the food chain and increases the yield of fish. Conversely at the centres of oceans smaller phytoplankton and planktonic bacteria grow in the nutrient-poor waters and are eaten by small herbivorous plankton. This lengthens the food chain and reduces its overall efficiency because food is transferred through more trophic levels. The final effect is the very low fish production characteristic of oceanic food chains.

The length and complexity of oceanic food chains and food webs may also be related to greater environmental stability at the centre of the oceans.

Phytoplankton growth, light and depth, compensation and critical depths

In the sea, light intensity is rapidly reduced with increasing depth (Figure 3.1). Phytoplankton needs light energy for photosynthesis and growth, and so is limited to the upper 100 m of the sea (Figure 4.5); at this depth in the open ocean the light intensity is about 1% of its surface level. At very high light intensities near the surface photosynthesis may be inhibited. Some of the total organic material produced by photosynthesis (gross production) is used in respiration, during which organic material is broken down. The remainder is termed net production (Figure 4.5A). Gross production falls with depth as light intensity falls, while respiration remains constant. The two exactly balance at the compensation depth. In the euphotic zone above this level (Figure 4.5A), there is a net gain of plant material, and production of organic material by photosynthesis exceeds breakdown of organic material by respiration. In the aphotic zone below the compensation depth there is a net loss of organic material, since breakdown exceeds production. At the critical depth the total production of organic material by photosynthesis in the water column above exactly balances the total breakdown by respiration.

Clearly since incident solar radiation changes with time of day and season, the compensation depth will move up and down in accordance, and the net production will vary. In the tropics, for example, the noon

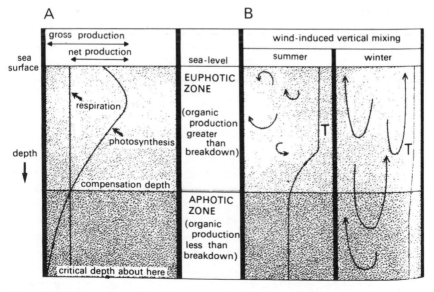

Figure 4.5 *A.* The relationship between depth and photosynthesis and respiration by phytoplankton.
B. Wind induced vertical mixing and the thermocline (*T*) during summer and winter in temperate climates.

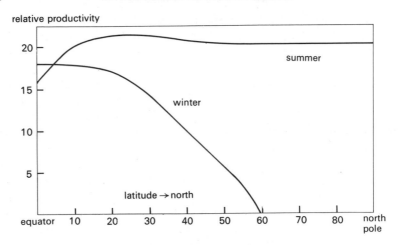

Figure 4.6 Summer and winter productivity in relation to latitude in the northern hemisphere. In summer relative productivity is similar in the tropics and polar regions, while in winter it is high in the tropics but almost zero at 60° N (modified from Hill, vol. 2, 1963).

compensation depth may be 100 m, while in temperate climates it is between 20 and 50 m in summer and zero in winter when there is no net production. In summer there may be little difference in relative productivity between the tropics and the poles, but in winter there is a very marked difference for obvious reasons (Figure 4.6). The phytoplankton themselves also absorb light, and as a result the compensation depth may rise as they increase in numbers.

Seasonal plankton cycle, grazing and vertical mixing

In temperate climates in summer, mixing of the surface water by wind action is shallow and is above the well defined thermocline and the compensation depth. In winter, stronger winds mix the water below the compensation depth and often below the critical depth, and the thermocline is destroyed (Figure 4.5B). This has a number of effects. In summer, phytoplankton are not carried out of the euphotic zone. They grow rapidly as soon as the thermocline is established, and soon deplete the nutrients (nitrate and phosphate) (Figure 4.7—North Atlantic, temperate). Nutrients are not replenished from below because there is little deep mixing, and so phytoplankton growth is soon inhibited. In autumn, wind mixing of the surface water increases, nutrients are brought up from deep water, and there is a second transient increase in phytoplankton. But this second peak

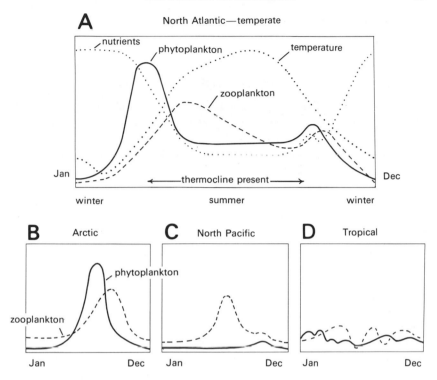

Figure 4.7 Geographical variation in the seasonal cycle of phytoplankton and zooplankton near the surface of the sea.

A: The cycle in North Atlantic temperature waters. Nutrients, temperature, and the presence of the thermocline are also shown on this diagram.

B, C, D: Less well known cycles in other areas. The vertical scales are arbitrary, but can loosely be read as phytoplankton and zooplankton biomass, °C, and nutrient (PO_4, NO_3) concentrations (modified from Heinrich 1962, and Parsons, Takahashi and Hargrave, 1984).

quickly decays as phytoplankton are taken into deeper water by more vigorous mixing and hence die. Zooplankton graze on the phytoplankton, and so a peak in zooplankton numbers follows each of the two phytoplankton peaks and will act with nutrient depletion to reduce the first peak, and with vertical mixing to reduce the second peak.

The spring phytoplankton decline is not always caused by zooplankton grazing and its effect varies from place to place. Inshore waters show great variability in light, temperature, turbidity and so on. Here, phytoplankton numbers need not be controlled by grazing. For example in Long Island Sound (New York, USA) phytoplankton numbers decrease in spring because of low nutrient levels. However in more oceanic conditions zooplankton grazing is usually the dominant factor.

Geographical variations in seasonal cycles and productivity

There are wide geographical variations in these relationships (Figure 4.7*B*, *C*, *D*). In the Arctic there is one phytoplankton peak followed by a zooplankton one, in the North Pacific *Calanus* nauplii in the zooplankton graze down phytoplankton before any spring phytoplankton peak can develop. In the tropics there are no peaks, and there is little relationship between phytoplankton and zooplankton grazing, except for some small fluctuations controlled by local weather.

In general, productivity of ocean waters is high in the Antarctic and Arctic, off the west coasts of America and Africa, and sometimes at the equator, where nutrient-rich deep water wells up. It is poor in the temperate climate centres of the North and South Pacific and Atlantic Oceans, and in the Indian Ocean (Figure 4.8).

Zooplankton feeding

The grazing effects of herbivorous and carnivorous zooplankton on phytoplankton depend on food density, size and shape, and on methods of capture.

Particle feeders such as copepods meet a wide range of particles whose numbers vary in space and time. Their feeding strategies are adapted to this. Some species feed selectively on the most abundant size peak but also prefer larger particles. Others can alter their feeding from small to large particles. In *Calanus pacificus*, a herbivorous species, feeding increases with diatom density to about 3000 diatom cell ml^{-1} and then remains constant as the density increases further (Frost, 1972). Below 3000 cells ml^{-1}, *C. pacificus* eats a greater proportion of bigger diatom cells. There are also intraspecific differences between individuals of different sexes and ages. The exact method of capture and particle size assessment is not understood, and involves considering viscosity and Reynolds numbers.

Little is known about the defence mechanisms of phytoplankton to grazing. For example, how do the large spines of phyto- and zooplankton affect their catchability, and what about the toxicity of many blue-green algae and dinoflagellates?

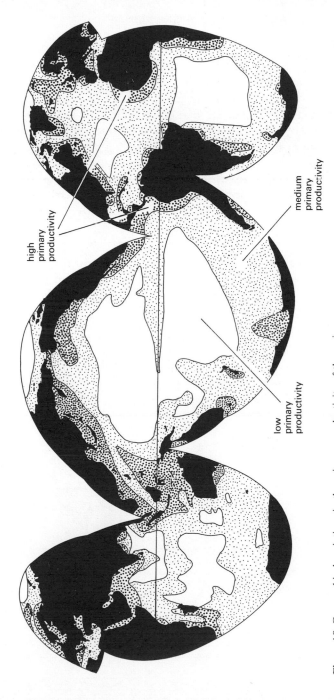

Figure 4.8 Geographical variations in the primary productivity of the major oceans.
No shading: areas of low productivity in the centres of oceans, primary productivity < 100 mg C/m^2/day.
Light shading: medium productivity 100–250 mg C m^{-2} d^{-1}.
Heavy shading: high productivity in the Antarctic and Arctic, and off the west coasts of America and Africa upwelling of nutrient-rich waters occurs; primary productivity > 250 mg C m^{-2} d^{-1} (modified from Koblentz-Mishke, Volkovinsky and Kabanova, 1970).

high
primary
productivity

medium
primary
productivity

low
primary
productivity

Plankton patchiness

Phytoplankton and zooplankton are patchily distributed on a large and small scale, both horizontally and vertically in the ocean (Figure 4.9). These effects are more marked in nearshore waters. Open ocean phytoplankton are less abundant and less patchy. Plankton patchiness is a complex phenomenon caused by differences in the physical, chemical, and biological nature of the environment with time. It can occur over a very wide range of scales from metres to kilometres, although zooplankton often occur in smaller patches than phytoplankton.

The large-scale ocean circulation systems move plankton vertically and horizontally, and there are many examples of invertebrate larvae being

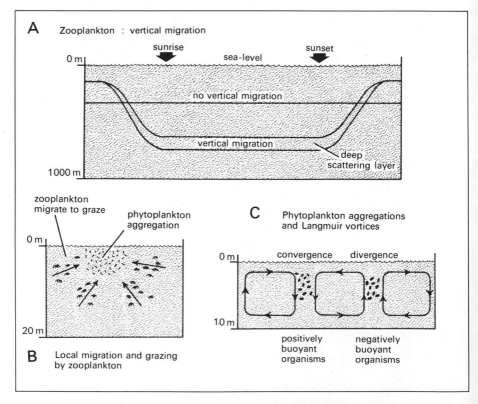

Figure 4.9 Patchiness of phytoplankton and zooplankton in the sea. The vertical scales of depth are approximate (modified from Bainbridge, 1957, Herring and Clarke, 1971, Langmuir, 1938, Stavn, 1971, Stommel, 1949).

carried across the Atlantic Ocean by the North Atlantic Drift (Scheltema 1971; Scheltema, in Brauer (ed.) 1972). On a small scale (Figure 4.9*B*) zooplankton may migrate into dense phytoplankton patches to graze, and then migrate out again. Langmuir vortices (Figure 4.9*C*) induced by wind action cause local convergences at the surface above downwellings and local divergences above upwellings. These vortices are on a scale of 1 to 10 m. Positively buoyant phytoplankton are likely to aggregate in the downwellings and negatively buoyant phytoplankton in the upwellings. The wind action producing the vortices often arranges them in long lines or windrows so that the phytoplankton are aggregated into long streaks which are clearly visible from a ship.

Vertical patchiness is caused by a number of factors. Phytoplankton growth is sometimes inhibited by too much light near the surface, or by too little light when growth of populations above causes shading. Many zooplankton migrate downwards in the water column before sunrise and upwards after sunset (Figure 4.9*A*). Echo sounding shows that different deep scattering layers may migrate together, while some remain at a constant depth. Photographs, net catches and observations from submersibles, have shown that these migrating layers are made up of planktonic crustaceans, siphonophores, and small fish. The mesopelagic myctophid fish *Cyclothone pseudopallida*, about 7 mm long, is amazingly abundant in some of the deep scattering layers. Some medium-sized fish also migrate vertically. The drift net fishery for herring off Scotland used to be based on the vertical migration of herring to the sea surface at night, and the nets were shot at this time (Figure 9.6).

Vertical migration is mainly controlled by behavioural responses to light and pressure. It probably enables zooplankton and fish to avoid surface predators during the day and also to graze different populations of prey by moving in one direction with the surface current at night and in another with the mid-water currents during the day. Currents in the upper layers of the sea often move more quickly than those at depth. Migration upwards at night would therefore allow faster lateral transportation at that time. It is also energetically advantageous to be in colder deeper water during the day. Food material may also be transported to great depths by a series of overlapping vertical migrations in the epi- and mesopelagic zones.

Differences in swimming and sinking rates are additional factors that will cause patchiness. Sinking rates of phyto- and zooplankton differ markedly (Table 4.1), and some larger phytoplankton such as the dinoflagellate *Gonyaulax polyedra* can swim at 2–20 m d^{-1}, which rivals the swimming rates of many smaller zooplankton.

Table 4.1 Sinking rates of phyto- and zooplankton $(m\,d^{-1})$ (Smayda, 1970).

	$m\,d^{-1}$		$m\,d^{-1}$
Living phytoplankton	0–30	Chaetognatha	~ 435
Dead phytoplankton	~ 1–510	Amphipoda	~ 875
Foraminifera	30–4800	Copepoda	36–720
Radiolaria	~ 350	Pteropoda	760–2270

Plankton indicator species

Many planktonic species fall into recognisable groups which delineate water masses that are difficult to identify physically or chemically. For example, diatoms are characteristic of polar waters, and dinoflagellates, coccolithophorids and blue-green algae are characteristic of tropical waters. Sometimes particular species are useful in this way. In the Pacific, the euphausiid *Nematoscelis gracilis* occurs in equatorial waters, *Euphausia brevis* occurs in waters from about 10 to 35° N and S, and *Thysanoessa gregaria* in waters from about 35 to 45° N and S. Around Britain, various species of chaetognaths and associated species enable one to differentiate between oceanic waters and inshore (neritic) waters. *Sagitta setosa* with the copepods *Temora longicornis* and *Centropages hamatus* and the diatoms *Biddulphia sinensis* and *Asterionella japonica* are characteristic of the southern North Sea and eastern English Channel. These inshore waters have slightly lower salinities and more widely fluctuating temperatures than the oceanic waters of the North Atlantic. *Sagitta elegans* with the trachymedusan *Aglantha digitale*, the copepod *Centropages typicus*, the euphausiid *Meganyctiphanes norvegica*, and the polychaete *Tomopteris helgolandica*, are found in the northern North Sea, in the waters off the west coasts of Scotland and Ireland, and in the western oceanic entrance to the English Channel. These *Sagitta elegans* waters are a mixture of oceanic waters derived from the North Atlantic Drift current, and from the inshore waters around the British Isles, and have a higher nutrient content than the true inshore *Sagitta setosa* waters.

Neuston

Neuston may be defined as very small planktonic organisms living within 1 mm and often within 100 μm of the sea surface. This thin film of water contains many biochemicals, detritus, surface-active substances, highly pigmented bacteria, and nanoplankton, and its constituents are qualitatively different from the sea water immediately underlying it (10–20 cm). Bacterio-neuston increase with increased wave and wind action. Euneustonic organisms always remain in the film, while facultative neuston enter it during darkness. The total neustonic biomass is very small. Neuston is sampled with a skimmer net on a frame towed along the surface.

Pleuston

Pleustonic organisms float at the water surface. The Portuguese man-of-war *Physalia*, a siphonophore coelenterate, is buoyed by a gas-filled float; its blue tentacles may hang down 30 m, and catch nekton prey. Other siphonophores with floats are also found at the water surface.

Epipelagic organisms, nekton

Epipelagic organisms, living in the top 200 m, are very diverse. We have already discussed the phyto- and zooplankton. Many nektonic species, such as squid, swordfish, tuna, and salmon, and the pelagic food fish of the continental shelf (herring, mackerel) live in this zone (Figure 9.4). Many epipelagic *nekton* (fish and squid) live in the tropics or subtropics and some migrate into temperate waters during the breeding season (great white and mako shark, tunny and swordfish). Many live on large prey (squid and small fish) while a few such as the whale shark and basking shark eat zooplankton. Copepods often make up 70 to 90% of a zooplankton sample in the epipelagic zone, and salps and doliolid tunicates may occur in large numbers. Pelagic polychaetes like *Tomopteris* are voracious carnivores with big eyes and jaws, while some bottom-dwelling polychaetes have a swarming epipelagic reproductive phase, the best known example of which is the palolo worm. Some of these species are luminescent. *Euphausia superba* (krill) is abundant in the Antarctic and is the main food of the baleen whales in that region. A considerable amount is known about the physiology and behaviour of some of the forms that live in the epipelagic zone. For example, *Euphausia* is known to contain high levels of vitamin A which may account for the high content of this vitamin in many fish livers; *Euphausia* also accumulates radioactive fall-out as ^{65}Zn. The purple gastropod *Janthina* (2 cm long) which preys on the siphonophore *Velella*, secretes a fluid which it wraps around bubbles of air to make its float. Finally, the amphipod *Phronima* (3 cm long) eats the tunicate *Pyrosoma*, and then proceeds to live in the tunicate's empty barrel-shaped test. These examples serve to illustrate the wide diversity of information that is available about organisms living in this zone.

Mesopelagic, bathypelagic, and abyssopelagic animals

Animals become less common but more bizarre in the *meso-, bathy-,* and *abyssopelagic* zones. Many forms are small, black, red, or transparent, and sometimes luminous (Figure 4.10).

In the upper 500 m fish are silvery or grey with their upper surface dark, and their sides and lower surfaces pale. At these depths light is still directional, and so it is of advantage to be countershaded in this way. When viewed from above a fish will appear dark against a dark background, while from below it will appear light against a light background. Silvery fish living well below these depths are not countershaded, presumably because the very low levels of light are non-directional (Figure 3.1). Many mesopelagic fish (100–1000 m) have large forward or upward-looking tubular eyes set in parallel to give wide angle binocular vision, while in some bathypelagic groups, such as the abundant Gonostomatidae (*Cyclothone*) and Myctophidae (lantern fishes), species with well developed luminescent organs also

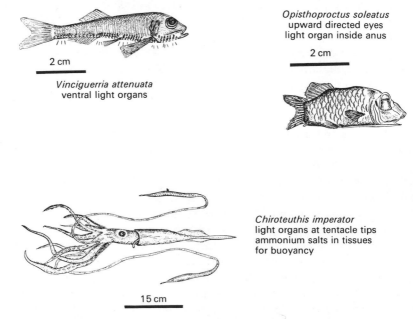

2 cm

Vinciguerria attenuata
ventral light organs

Opisthoproctus soleatus
upward directed eyes
light organ inside anus

2 cm

Chiroteuthis imperator
light organs at tentacle tips
ammonium salts in tissues
for buoyancy

15 cm

Figure 4.10 Bathypelagic animals (modified from Herring and Clarke, 1971).

have large eyes. In species with poor vision, olfactory organs and lateral line papillae may be well developed. Many of these deep-water fish are also very small, being only a few centimetres long.

Mesopelagic luminescent squid and fish often migrate to the surface at night, while darker mesopelagic fishes are powerful predators with large teeth able to eat animals of their own size. Luminescent jelly fishes, tunicates, copepods, shrimp and prawn are all mesopelagic, and luminescence may be important in species' recognition, sex and feeding.

The chaetognath *Eukrohnia hamata* lives at 0–400 m at 60° N and S, but at 800–1200 m in the tropics. Many mesopelagic animals are distributed similarly. In some groups species at different depths are different in size. Calanoid copepods are 1–2 mm if epipelagic, reach 17 mm if meso- and bathypelagic (500–2000 m), and below that depth get smaller again. Many squids live in deeper water though little is known of their biology. Some produce a luminous shower of sparks, while others such as *Chiroteuthis imperator* (27 cm) have light organs on the end of their tentacles which may be food-catching devices (Figure 4.10).

Pressure in the deep ocean is very high, reaching 1000 atm at 10 000 m (Chapter 3). The biochemical and physiological effects of these enormous pressures on deep-sea organisms appear to be complex (Macdonald, 1975). Cytoplasmic structures (e.g. microtubules), cell transport, the metabolism of DNA, and cell excitability are all sensitive to pressure, for example.

Teleost fish have evolved a pressure-sensitive organ, the swim bladder (or air bladder) which relics for its function on the inverse relationship between the volume and pressure of a gas. They can regulate the volume of their swim bladder, which is gas-filled, and hence are able to adjust their buoyancy to be neutral in relation to the ambient pressure. As a fish moves downwards its swim bladder in compressed by the increasing external pressure and the fish has to secrete gas into the bladder cavity in order to maintain neutral buoyancy. Conversely, as the fish moves upwards the volume of its swim bladder increases and it resorbs gas. The gas is a mixture of oxygen (which predominates), nitrogen, carbon dioxide, and small quantities of argon and neon. The proportions vary between different species.

Many pelagic teleosts that live in the open ocean, such as tunny, flying fish, herring, and mackerel, have well developed swim bladders, and so can maintain their position in mid-water easily; in this way more energy is likely to be available for horizontal locomotion. Other teleosts, such as the flat fishes, blennies, gobies and weaver fish, all of which as adults live on the sea bottom, have no swim bladder, and so like all elasmobranchs must swim actively to keep in mid-water. Amongst the deep-water pelagic teleosts, the gonostomatids and myctophids, which live between about 100 and 500 m have swim bladders, while some groups such as the ceratioid angler fish, the gulpers (*Lyomeri*) and the Alepocephalidae, that live below 500 m as adults, have lost their swim bladders. There is no obvious reason for this relationship with depth, and there are a number of exceptions.

In general, the bathypelagic zone (1000–4000 m) is cold (0–5° C) and totally dark except for bioluminescence. Biomass is about 10% of the mesopelagic zone, and there is a characteristic community of invertebrates and vertebrates. Some are important sources of food for other species. *Vinciguerria attenuata* (5.6 cm) (Figure 4.10) is a bathypelagic gonostomatid fish related to *Cyclothone*. It is very common in the Pacific at depths greater than 1000 m and is a recognised food source of tuna and albacores. *Opisthoproctus soleatus* has the unusual distinction of possessing a light organ inside its anus which shines the light along a channel onto a dorsal reflector that reflects the light downwards between its ventral scales (Figure 4.10).

Deep water sampling methods

Most of these species are caught with specially designed nets or photographed from submersibles. The Isaac Kidd mid-water trawl has a large mouth and a very long net, and is towed at about 500–2000 m depth, while the double net designed at the Institute of Oceanographic Sciences, Godalming, England, is triggered electronically to open and close at set depths (Figure 4.11).

Side-scan rader is now used to study objects in a range of depths (Belderson, Kenyon, Stride and Stubbs, 1972; D'Olier, in Dyer (ed.) 1979). Regular short sound pulses of about 1 m sec are transmitted laterally from a transducer sited beneath a research vessel, and echoes from underwater objects (fish shoals, sediments, rocks) are received by the transducer. A paper print-out, or *sonograph*, provides a permanent record of the objects, although this is not easy to interpret at first sight. In an instrument with a 1 km range, the main beam of the sound pulse usually has a horizontal spread of 2° and a vertical spread of 10°; secondary beams give added information. Resolution is about 7 m at 22 km and about 15 cm at 300 m. The equipment has been used to study the local topography of the sea bed in the deep sea, on the continental shelf, and in harbours and estuaries. It has also been used for wreck identification, for investigations on vertical

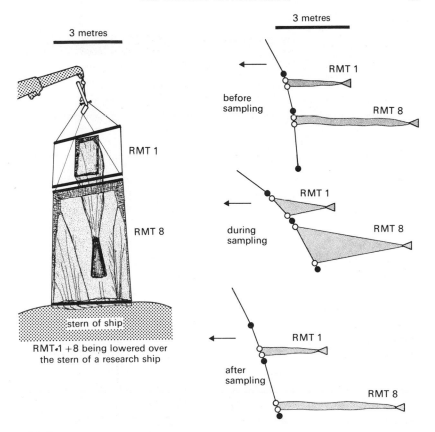

Figure 4.11 The NIO double-closing combination net (RMT 1 + 8) for deepwater fishing.
Left: the nets being lowered over the stern of a research vessel.
Right: diagrams of the nets before, during, and after sampling at depth (modified from Baker, Clarke, and Harris, 1973).
NIO = National Institute of Oceanography, England, now renamed IOS, the Institute of Oceanographic Sciences.

migration, and in fisheries biology to detect fish shoals, spawning grounds and fishing gear.

Whales, migration, krill

Whales are marine air-breathing mammals and are probably the most intelligent animals apart from man. The toothed whales, suborder Odonto-

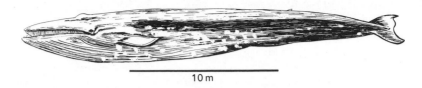

10 m

Figure 4.12 The blue whale *Balaenoptera musculus*.

ceti, include the common porpoise (*Phocaena phocaena*), the killer whale (*Orcinus orca*), and sperm whale (*Physeter catodon*). The sperm whale grows to 20 m and can stay under water for over an hour. It is the deepest diving whale often diving below 1000 m, and eats squid—some of which can be over 9 m long.

The baleen whales, suborder Mystacoceti, include the largest whales: the Blue Whale (*Balaenoptera musculus*), the Fin Whale (*Balaenoptera physalus*), the Sei Whale (*Balaenoptera borealis*), and the Southern Right Whale (*Eubalaena glacialis*). The baleen whales have mouths, and feed on plankton (krill and copepods) and some times on small fish. They catch their food by filtering water through a sieve of whalebone plates (baleen) that hangs from the roof of their mouth. The Blue Whale is the largest animal to have lived on earth (Figure 4.12). It can be 25 m long and lives for over 50 years. Calves are 7 m at birth and adults eat about 3 tonnes of krill per day.

The baleen whales live in temperate and cold waters, especially in the Antarctic. In winter they migrate from their feeding grounds in the Antarctic towards warmer equatorial waters where they reproduce and nurture their young. The Southern Right whale makes an annual 7400 km migration from its breeding grounds off Argentina to Antarctica. To achieve this they have to cruise at about 6 knots (*c*. 11 kph).

Sea birds; feeding, migration

Sea birds exploit most of the living organisms at and near the sea surface. They eat fish, crustaceans, plankton and squid, and do so by diving, scavenging and pursuit of bird by bird.

Sea birds are most common near or on coasts and above upwellings since food is most abundant there. They nest on land or on the shore—shearwaters in burrows, albatrosses on grassy areas, terns on sandy shores or dunes. Most of them breed in very dense colonies containing thousands of individuals, although there are exceptions: the flightless cormorant forms breeding colonies of about 20 birds.

Sea birds fall into three broad ecological groups depending on their behaviour and feeding. Some spend most of their life over and on the ocean (the shearwaters and petrels); some are divers and catch fish below the sea surface (penguins and gannets); and some are coastline birds that sometimes move inland (the gulls, cormorants and coastal terns).

A number of species perform the most remarkable annual migrations to and from their breeding grounds. The Arctic Skua breeds in the Arctic and winters in the Antarctic, and the Arctic Tern undertakes a similar migration. The shearwaters wander all over the oceans, sometimes even circumnavigating the major ocean basins (Pacific: Short-tailed Shearwater, Sooty Shearwater; Atlantic: Great Shearwater, Sooty Shearwater).

CHAPTER FIVE

OCEAN BED FORMATION

A minimum potential energy earth

If the earth's surface was in a state of minimum potential energy, it would be smooth and covered by a uniform ocean of constant thickness. The sea floor would be 2.44 km below present sea level and covered by an ocean 2.64 km deep. The ocean surface would then be 2.64–2.44 = 0.2 km (200 m) above present sea level. The earth's surface is not at minimum potential energy however; it is concentrated at two levels. The continents and continental shelves are at + 0.4 km above sea level, and the deep ocean at − 4.5 km below sea level (Chapter 1, Figure 1.3).

Slow vertical movement of the earth's surface, isostasy

The various rock masses which make up the earth's crust behave rather like semi-solid liquids of high viscosity when considered on the scale of the earth. They move slowly up and down on, and also from place to place over, the earth's mantle. For example, 12 000 years ago, when the 2 km thick layer of ice over North America and Scandinavia melted, the weight removed was equivalent to one-fifth of the difference in height between the continents and the oceans. As a result, the land under the ice rose, rather like a spring with a weight on top of it if part of the weight were removed. Many ancient raised beaches show the effects of this rise in sea level. As an analogy, icebergs are slightly less dense than sea water, and float in the sea with most of their mass under water. The greater the mass under water, the greater the mass it can support above water; in other words if the mass of ice under water increases, the iceberg can sit further out of the water. This principle when applied to continental masses is called *isostasy*. It assumes that the less dense rocks of the earth's crust (about 33 km deep in the continents) are floating on the more dense mantle. It also implies that the higher land masses have deeper crustal foundations (like the iceberg).

Evidence from gravity anomalies, seismic reflection and refraction, and continuous reflection profiling

A study of gravity anomalies, of seismic reflection and refraction, and of continuous reflection profiling data shows that the principle of isostasy is largely correct. The average acceleration due to gravity on earth is $980 \, \text{cm/s}^2$ or 980 Gal, but varies from 978.04 at the equator to 983.22 at the poles because of the flattening of the earth at the poles and the centrifugal force of rotation of the earth. It also decreases $0.001 \, \text{Gal km}^{-1}$ ($1 \, \text{mGal km}^{-1}$) outwards from the earth's centre, and varies with the surface topography of the earth—for example it rises near a high mountain. If all these effects are allowed for, and the average gravity calculated for an earth where continents are levelled to sea level and where the ocean is filled with rocks of the same density as the continents, the observed gravity values do not agree with those expected. (Gravity values are obtained from an accurate gravity meter carried by a ship or aircraft.) The observed values are about 100 mGal below average over the continents and about 100 mGal above average over the deep oceans. These gravity differences or anomalies suggest that the less dense crustal rocks are deeper under the continents than under the sea.

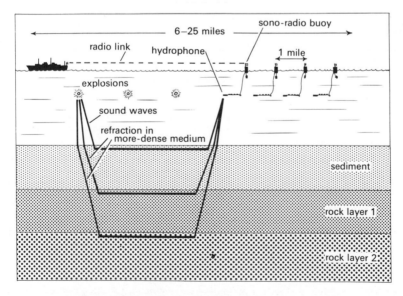

Figure 5.1 Seismic refraction (modified from Hill, vol. 3, 1963).

Seismic reflection and refraction measurements are made as follows (Figure 5.1). A research vessel releases a small explosive charge. The sound from the explosion travels through the water column and into the different layers of the earth's crust below the ocean bed. Some sound is reflected at each interface—water/rock, rock layer 1/rock layer 2—and some is refracted at each interface. The reflected and refracted sounds are picked up by hydrophones attached to buoys which have been released by the same or a companion research vessel. Information from these hydrophones is transmitted from the buoys to the ship by radio. Sound velocity increases with the density of the material, and so reflected and refracted waves will return faster through the denser deeper rock layers. The differences in times of arrival of sound at the detector hydrophones enable one to calculate the densities and depths of different rock layers. In practice, four buoys are used at one mile intervals. The explosives are set by a time fuse, and detonate at 100 m depth, the first about 6 km from the nearest buoy then at 4–8 per hour. Ten to twenty explosions are fired for a full profile with the ship steaming at about 8 knots. Recently other energy sources, such as electric sparks, compressed air and propane have been used. The similar technique of continuous reflection profiling uses a continuous sound of between 20 and 150 Hz (cycles s^{-1}—a low pitched note), and shows clearly how uneven contours in the rock under the sea are filled with sediment (Figure 5.2).

Data from seismic refraction and reflection, and from continuous reflection profiling, confirm the conclusions from gravity anomalies

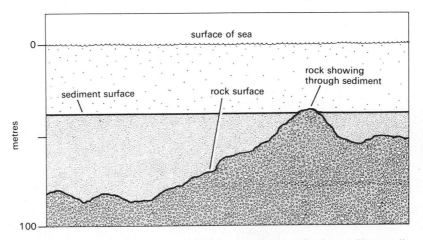

Figure 5.2 Bottom and sub-bottom profile from a continuous reflection profile recording. Uneven rock contours are filled in by sediment (modified from Hill, vol. 3, 1963).

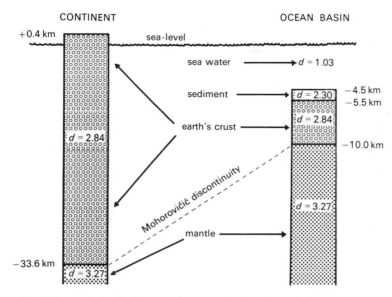

Figure 5.3 Difference in depth of the earth's crust on land and beneath the sea (modified from Weyl, 1970).

(Figure 5.3), and prove that the crust is considerably deeper under the continents than under the ocean basins. The Mohorovičić discontinuity, the line separating the earth's crust from the mantle, lies at 33.6 km under the continents and only 10 km under the oceans. If the total mass cm^{-2} under the continents (depth × density) is compared with that under the sea, taking into account the density and depth of sea water, and measuring to the same level of -33.6 km, the values are the same: 96.2×10^5 g cm^{-2} for the continents and 96.8×10^5 g cm^{-2} for the oceans—a remarkable agreement (Weyl, 1970, and data in Figure 5.3).

These techniques and calculations, then, show that the continents and ocean basins are equally balanced, that the high land masses have deeper crustal foundations, and that the principle of isostasy holds.

Precipitation and evaporation, continent erosion and isostatic adjustment

Continents receive more water by precipitation than they lose by evaporation, and the difference, about 10^9 kg s^{-1}, is returned to the oceans by the rivers. Rivers continuously erode and transport sediments to the sea, and therefore are continuously reducing the difference in height of the continents and ocean basins by eroding from the former and depositing on the latter. Weyl (1970) has calculated that with present river flow rates and sediment content, it would need only 11 million years to erode the United States to sea level: in other words, it would have been eroded sixty times since the beginning of the Cambrian period. Clearly other forces must allow the continents to remain at about their present level.

Isostatic adjustment reduces but does not completely eliminate the effect of this erosion. Removing sediment from the land and placing it on the sea bottom (like moving weights on a balance from one side to the other) will let the land rise and make the sea bed fall. In practice, 3.5 km of a continent must be eroded to lower its level relative to sea level by 1 km. Taking this into account, the United States would be eroded to sea level in 40 million not 11 million years. But even this does not explain how the continents have remained above sea level since the beginning of the Cambrian Period. Furthermore if erosion had been continuous since then, the ocean floors should be covered by 30 km of sediment or sedimentary rocks. This is not so. The sediments are about 1 km thick, and the earth's crust 4.5 km thick under the deep ocean. The explanation comes from recent research showing that the continents and ocean bottoms are in a state of continuous change, and that new sea floor is being formed and destroyed all the time.

Sea floor spreading, plate tectonics

The earth's surface is divided into a number of large flat building blocks or tectonic plates which are moving relative to each other. When they move apart new crust is formed; when they come together either mountain ranges fold up, or one block dips below the other, thus forming the ocean trenches. These sites are often centres of earthquakes and volcanic activity. Movement of the blocks is probably caused by huge convection currents within the mantle which obtain their energy from radioactive decay within the earth. Let us look at the evidence for these statements.

Palaeomagnetism, continental drift, South America/Africa coastline fit

Firstly, all the continents contain pre-Cambrian rocks and Cambrian sediments containing fossils from 500 to 600 million years ago, and so the continents can be regarded as permanent features of the earth's crust, at least over this time scale. Palaeomagnetic data from sedimentary and igneous rocks, however, indicate that their position has changed greatly. Small particles of a natural magnetic rock containing the iron ore magnetite

(Fe_3O_4) are common in sediments. These particles point freely to the poles when suspended in water, but retain that direction when buried in sediments, thus indicating the position of the poles at the time at which they were buried. Similar information is available from igneous rocks. As blocks of igneous rock cool, some minerals (e.g. magnetite) become magnetic below about 500 °C—their *Curie* temperature—and retain the magnetic direction imposed upon them at the time of their solidification. The magnetism retained in rocks in these ways is very weak, but can be detected with suitable instruments, and indicates the past positions of the poles. For example, the North Pole measured from European and North American rocks was in the middle of the Pacific Ocean 500 million years ago, and has moved towards its present position since then. However, the paths taken by the poles are different when measured from European and North American rocks. The two continents, therefore, have moved relative to the North Pole and relative to each other, a movement called *continental drift*. All the earth's continents have moved in this way.

Other evidence leads to the same conclusions. The east coast of South America fits well with the west coast of Africa and the geological features of both match almost perfectly (rocks, mountain ranges). Fossils formed on the two continents before the Cretaceous Period (about 120 million years ago) are very similar, while more recent ones are different. Africa and South America were joined before the Cretaceous Period, therefore, and have drifted apart since then forming the Atlantic Ocean between them in the process.

Mid-oceanic ridges, magnetic reversal and sea floor spreading, *Glomar Challenger*, Deep Sea Drilling Project

As continents separate, new ocean floor is produced at the mid-oceanic ridges in the centre of both the Pacific and Atlantic Oceans and is spreading outwards from them in both directions (Figure 5.4). The evidence for this is as follows. During the recent magnetic history of the earth the North and South Poles have reversed a number of times. If new ocean floor is being continuously formed at the mid-oceanic ridges, the magnetic elements in the newly formed crustal rocks should take up the magnetic direction of the poles at that time. This would be caused by rocks cooling through their Curie temperature as they come to the surface at the mid-Atlantic ridges, and would result in strips of reversed magnetism as one moves away from the ridges in either direction. Magnetometer records from research vessels travelling across the ridges show exactly this (Figure 5.4), and indicate that

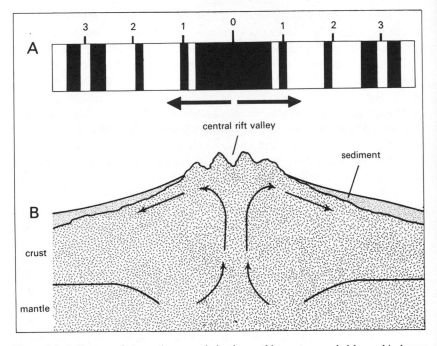

Figure 5.4 *A*. Pattern of magnetic reversals in the earth's crust, recorded by a ship-borne magnetometer across an oceanic ridge. Figures are millions of years before present.
B. Formation of new oceanic floor at the mid-oceanic ridge. Note that sediment is thinner near the ridge where new crust has just come to the surface (modified from Weyl, 1970).

the sea floor is spreading at about 2 to 5 cm per year. The American-financed Deep Sea Drilling Project (DSDP) is operating the research vessel *Glomar Challenger* which carries sophisticated deep-sea drilling equipment. The vessel has operated since 1968, and is producing results that are comparable in importance to those of the British *Challenger* Expedition in the late nineteenth century. Among other results, drilling over and near the mid-oceanic ridge has shown that the sediments are young compared with sediments from other parts of the ocean. Sediments are very thin near the ridges, since the ocean floor has been exposed for a short time there, and become deeper as one moves away from them. These observations substantiate theories of sea floor spreading from the mid-oceanic ridges.

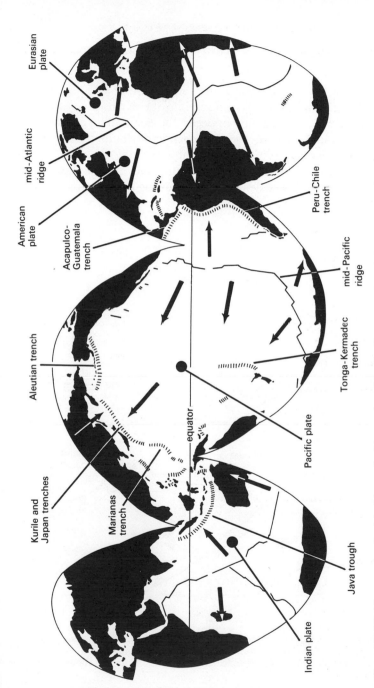

Figure 5.5 World oceans showing (*a*) mid-oceanic ridges — the continuous lines approximately in the centres of the Atlantic, Indian and Pacific Oceans, (*b*) formation of new ocean floor spreading outwards from the mid-oceanic ridges (arrows), (*c*) ocean trenches where one plate is dipping beneath another (rows of palisade lines), (*d*) major tectonic plates (some are named and indicated by a solid black circle in their centre). The ocean trenches and the mid-oceanic ridges are centres of intense earthquake and volcanic activity.

Lines of earthquake and volcanic activity, ocean trench and mountain formation

The creation of new ocean floor along the mid-oceanic ridges must be balanced by the compression or sinking into the earth of an equivalent surface area elsewhere. If this did not occur, the earth's diameter would be expanding at about 2 cm/year. The compression or sinking causes stresses and strains at the surface, resulting in earthquakes and volcanoes. As is to be expected, therefore, the distribution of earthquakes and volcanoes on the earth's surface lies along well defined lines following the mid-oceanic ridges (Figure 5.5). They also lie:

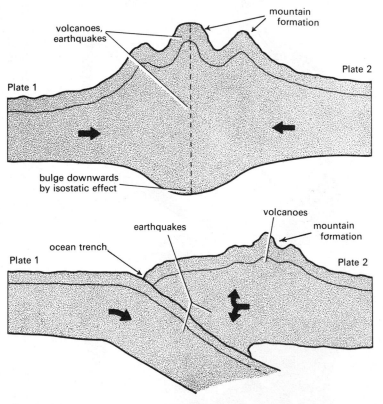

Figure 5.6 *A*. Diagrams of mountain formation. *B*. ocean trench formation where two tectonic plates push together. Mountains are formed when the plates approach at 5–6 cm yr^{-1}, and trenches and mountains when they approach at 6–9 cm yr^{-1}. Note that mountains can be formed in both instances.

(i) on some continents where ocean ridge-like structures have formed, like the East African ridge and Rift Valley system (Lake Tanganyika and Lake Malawi)

(ii) along the edges of some continents with high mountains and ocean trenches nearby (Peru and Chile and the Peru-Chile Trench, Mexico and Guatemala and the Acapulco-Guatemala Trench)

(iii) along lines of oceanic volcanic islands with an ocean trench near them (Aleutian Islands and Trench, Philippine Islands and Trench, Marianas Trench and Islands (Guam)). The Pacific Ocean is in fact ringed by ocean trenches (Figure 5.5).

Formation and destruction of the earth's crust

The mid-oceanic ridges, and examples such as the East African Rift Valley system, are places where new crust is being added to the tectonic plates. The edges of continents near mountain ranges, and the volcanic islands and ocean trenches, are places where surface area is being lost. Adjacent plates push against each other and buckle (mountains) or dip beneath each other (ocean trenches) (Figure 5.6). Buckling occurs when plates approach each other at 5 to 6 cm per year, and dipping takes place when they approach at 6 to 9 cm per year. At intermediate speeds dipping and buckling may occur together.

The creation of new ocean floor by sea floor spreading from the mid-oceanic ridges is therefore balanced by its removal at island arcs, continental edges, and ocean trenches where the tectonic plates crumple or sink. The movement of continents in relation to each other (continental drift) is part of the same process.

Continental shelves and dams, ice age sea level

The geophysical techniques of seismic refraction and reflection, cores from drilling, and precision depth recordings (PDR), have given us a detailed picture of *continental shelf formation*. Most continental shelves are underlain by long prisms of sedimentary rocks that stretch away from the coastline, while a few are underlain by igneous and metamorphic rocks. Natural rock dams hold many sedimentary deposits against the continents, and sediment is then deposited behind them. There are three types of dams. Tectonic dams are formed by geological upwelling or by volcanic lava (west coast of North America), reef dams are produced by marine corals or algae (Great Barrier Reef, Australia), and diapir dams are caused by slow upwelling of lighter sediments (*salt domes*) through denser rocks (Gulf Coast of North America). The picture may be complicated by huge underwater landslides (*slumping*), geological faults causing a sudden change in the depth of the shelf bottom, and by the effect of the four retreats and advances of the sea level during the four Ice Ages. The effect of the Ice Ages can be seen in the submerged beaches and terraces, and the channels of Ice Age rivers, which are found on many shelves. Continental shelf sediments also contain freshwater peat, mammoth bones and early human tools. The most recent sea level change began 35 000 years ago when the sea was at its present level. The sea level then slowly fell to about 150 m below its present level 15 000 years ago and than rose again to its present level about 5000 years ago. Changes like this will have exposed large parts of the continental shelves and greatly modified them.

CHAPTER SIX

THE BENTHIC ENVIRONMENT

This chapter is divided into two parts. The first describes the sedimentary environment and covers the formation, distribution and properties of seabed sediments, manganese nodules, sediment diagenesis and sediment-ation. The second part describes the benthic fauna and flora, and covers a number of topics. These include the macro- and meiofauna, bioturbation, macro-algae and macro-algal production, benthic communities and larval biology, community diversity and stability, the effects of increasing water depth, adaptations of deep sea animals, and hydrothermal vents.

A. The sedimentary environment

Continental shelf sediments, bottom currents

Sediments on continental shelves are all terrigenous, being derived directly from the land (Figures 6.1, 6.3). Emery (1968) divides them into relict sediments (70%) and modern sediments (30%). The former were laid down during the sea level changes of the Ice Ages. Modern sediments are being formed at the present time and are mainly produced *in situ* or are carried to their present sites by depositional agencies (wind, ice rafting, water currents).

Emery distinguishes the following sediment types: detrital (laid down by water, wind, ice—the typical outwash from a continent), biogenic (from animal and plant carbonate shells and tests), volcanic (debris near volcanoes), authigenic (minerals, such as phosphorite and glauconite that come out of solution under suitable conditions), and residual sediments (produced by *in situ* weathering of bedrock). The position of these sediments on the shelves around a typical ocean is influenced by the main oceanic gyres and their associated temperature differences (Figure 6.1). In the northern hemisphere the clockwise current flow produces warmer condi-tions on the western side up to a certain latitude, and biogenic sediments

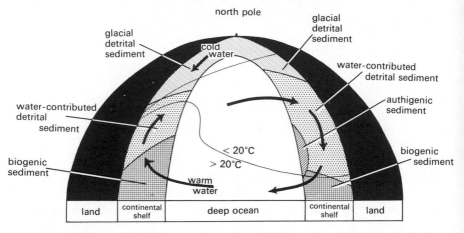

Figure 6.1 Sediment distribution on the continental shelves of an idealised ocean in the northern hemisphere (modified from Emery, 1969).

extend further north on that side. The effect is limited, however, by the southward counterflow of cold Arctic water. Seventy per cent of the sediments have been laid down after the rapid Ice Age fluctuations in sea level, that is within the last 15 000 years, and their main source is river sediment.

In the transition zone between MLWS and the continental shelf proper, down to 10 or 20 m (the maximum depth of wave influence), sediment grain size often decreases from sand into mud; these sediments are all terrigenous and most of them are derived from rivers or from beach erosion (see Chapter 7). On the shelf itself, sand, mud, and shell-gravel sediments predominate. Some continental shelves are mainly floored with sandy sediments, while others contain much mud. Shell sand is common at the outer edge of the continental shelf or on small submerged hills that are inaccessible to muddy sediments. Modern shelf mud deposits are a mixture of clay and silt which develops where currents are slow, and are deposited at a rate of 20 to 50 cm of sediment per 100 years (Heligoland, North Sea).

Present-day water currents near the sea bed on the continental shelf can be very slow and may not disturb the sediment, or can move at up to 100 to 150 cm s^{-1}, in the North Sea and English Channel for example and cause erosion. These latter currents are reduced by friction near the bottom, but are still strong enough to scour some areas. They also produce megaripples at the sediment surface that have a wavelength of 1 to 30 m and an amplitude of 0.6 to 1.5 m.

Bottom topography, abyssal plains and hills

The continental shelf changes to a slope at its outer edge, and then to a rise leading to the flat abyssal plains. Deep canyons lead from the shelf across the slope and rise. The plains merge into abyssal hills, are cut by ocean trenches, and are punctuated by underwater mountains and the mid-oceanic ridges. Most of these surfaces are covered by sediment. Some are scoured by bottom currents strong enough to produce ripples and to expose rock. Steep inclines are rare, the almost vertical sides of some ocean trenches being an exception. An accurate picture of bottom topography is given by the precision depth recorder (PDR), which is an echo sounder that

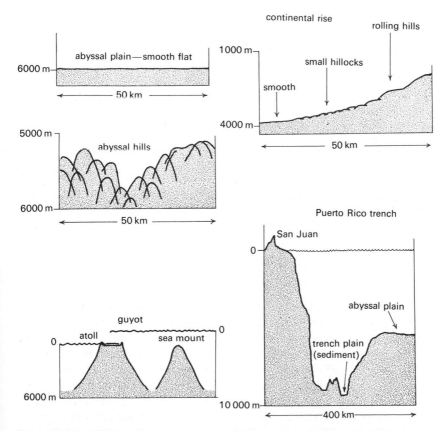

Figure 6.2 Topography of the deep sea as revealed by precision depth recording (PDR)—for explanation see text.

records detailed bottom topography by accurately measuring the time of a sound impulse travelling from a research vessel to the bottom and back again. The continental slope and rise may be as flat as the abyssal plains, show small hillocks, metres to tens of metres high, or have rolling hills (Figure 6.2). The flatness is partly caused by turbidity currents depositing an even layer of sediment (see below). The abyssal plains often merge at their seaward edge into ranges of abyssal hills, elevation about 10 to 500 m, and width 0.5 to 10 km. Ocean trenches often have steep and sometimes vertical sides, and vertical rock surfaces and overhangs have been seen from submersibles.

Sea mounts and guyots

Sea mounts and guyots interrupt the deep sea floor, particularly in the Pacific, and are caused by volcanic action. Guyots at one time reached the sea surface, and coral atolls formed on their tops (Figure 6.2). They then sank again. Sea mounts are usually more conical and have never been exposed to the atmosphere. Both have bare rock surfaces and often show evidence of strong scouring by currents.

Microtopography

Microtopography of deep sea sediments obtained by still and cine bottom photography from ships and deep sea submersibles can indicate bottom current conditions (scouring, and animal tracks and burrows) (Figure 6.15, p. 137). Measurements of sediment ripples and data from neutrally buoyant floats give bottom current speeds of $5-15\,\mathrm{cm\,s^{-1}}$.

Turbidity currents and sediment transport

Turbidity currents are intermittent currents of semi-fluid mud and sand dislodged from the continental shelf or slope by earthquake action. They transport large volumes of mud, sand, plant debris and so on from the continental shelves to the abyssal plains. Green leaves and grass have been collected at a depth of 4000 m from the abyssal plains in these areas, and so the turbidity currents must be relatively fast. The abyssal plains have a number of feeder channels or canyons down which the turbidity currents transport material. These channels run from the continental shelf to the abyssal plains and often occur where large rivers discharge into the sea (Mississippi, Hudson, Congo). One of the best known effects of a turbidity current occurred after the Grand Banks, Newfoundland, earthquake of 18th November 1929. The earthquake started a series of slumps and turbidity currents which broke about 13 telephone cables on the continental slope and rise, whose position and time of break gave speeds of 55 knots at first, slowing to 12 knots 640 km away on the abyssal plain. Subsequent coring showed that a 1 m thick layer of graded silt with shallow water micro-fossils had been deposited for 720 km over the abyssal plain. Many other turbidity currents are documented by similar data, and deep cores show bands of shallow water animal and plant detritus with poorly sorted terrigenous sand and mud from past currents, interspersed with normal pelagic sediments, such as brown pelagic clay. Terrigenous deposits from turbidity currents are not present on hills that rise above the abyssal plain, because the turbidity current flows around them or stops at them. For example, the abyssal plain between the continental shelf of North America and Bermuda contains terrigenous deposits from the continent, but the Bermuda Rise contains only pelagic clays.

Deep sea terrigenous and pelagic sediments

Sediments in the deep sea beyond the continental shelves come from a number of sources (Figure 6.3). They are broadly divided into *terrigenous sediments* carried to the sea from the land and *pelagic sediments* formed in the sea. They are thick (> 1000 m) under productive ocean areas such as polar regions and the narrow equatorial belt (Chapter 4), and thin (< 200 m) under oceans with a relatively low productivity, in the temperate zone for example. Terrigenous sediments come from the land and are carried by water currents, turbidity currents, and slumping, into the deep sea. Melting glaciers that release land debris as they melt also carry land material into the deep sea (rafted sediments). Terrigenous deposits are deposited over the continental shelves, slopes and rises, and on the nearby abyssal plains, and are relatively coarse grained.

Pelagic sediments are finer than terrigenous ones. *Brown* or *red clay* covers large areas of many ocean basins particularly in the Pacific. It is very fine (< 2 μm), and contains clay minerals originally derived from the land, volcanic ash and cosmic spherules from outer space. Its colour is due to oxidised iron, and it contains little material of biological (biogenic) origin. *Diagenic* deposits are made of minerals crystallising in sea water as authigenic minerals. Examples are *manganese nodules*, and *phillipsite*—a potassium rich silicate mineral with zeolite properties (ion exchange) common in Pacific deep sea sediments and nearby volcanic areas.

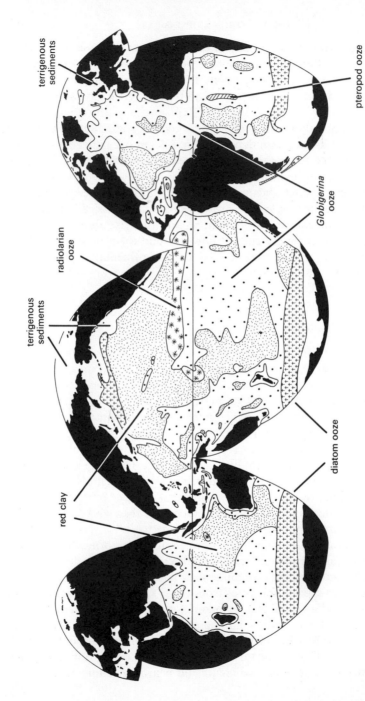

Figure 6.3 World distribution of pelagic sediments in the deep sea. The larger areas of each sediment type are labelled; other areas of that sediment can be identified by the same pattern of shading. Terrigenous sediments, which are unshaded, are near the continents and include all the sediments on the continental shelves (modified from Sverdrup, Johnson and Fleming, 1942).

terrigenous sediments

pteropod ooze

Globigerina ooze

radiolarian ooze

terrigenous sediments

diatom ooze

red clay

Biogenic pelagic sediments are defined as containing more than 30% skeletal material from dead planktonic organisms. This material gradually sinks through the water column and is carried from place to place by water currents as it does so. All naturally occurring oozes are a mixture of skeletons with one dominant type (> 50%).

Foraminiferan or *Globigerina ooze* and *coccolithophorid ooze* contain the $CaCO_3$ skeletons of the unicellular zooplanktonic protozoan *Globigerina* (Foraminifera) and the phytoplanktonic coccolithophores (Haptophyceae). *Pteropod ooze* contains the $CaCO_3$ shells of planktonic pteropod molluscs. The occurrence of these $CaCO_3$-rich sediments involves a complex story (Figure 6.4). $CaCO_3$ occurs naturally in two forms, calcite and aragonite. Calcite is less soluble than aragonite and has a higher density. *Globigerina* and coccolithophores contain calcite, and pteropods contain aragonite.

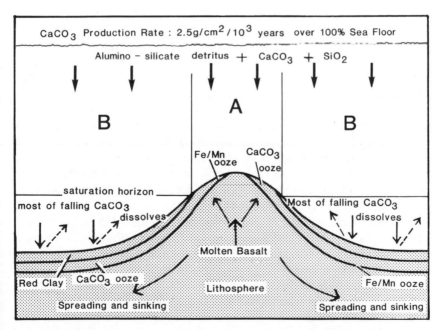

Figure 6.4 Calcium carbonate production by biological activity and the sedimentation of alumino-silicate detritus, calcium carbonate, and silica. Incorporation into sediments above and below the $CaCO_3$ saturation horizon ($CaCO_3$ compensation depth). *A*: 20% of the ocean sea floor is above the saturation horizon, and $CaCO_3$ accumulates in the sediments. *B*: 80% of the sea floor is below the saturation horizon, and most of the $CaCO_3$ dissolves before it reaches the sea bed (after Broecker, 1974).

Surface waters are supersaturated with both chemicals, and the solubility of both increase with decreasing temperature and with increasing pressure— in other words with depth. In particular, saturation drops rapidly across the thermocline.

In the Pacific, *aragonite* becomes undersaturated below 200 to 400 m, and *calcite* below 400 to 3500 m depending on the location. In the Atlantic the depths are 2000 m for aragonite and 4500 m for calcite. These depths are called the $CaCO_3$ *compensation depths, saturation depths* or *horizons*. The difference between the depths in the two oceans is because the Pacific is more acidic than the Atlantic—Pacific waters contains about 2.5 times the H^+ ion concentration of Atlantic waters.

Calcite and aragonite are not present in sediments below the saturation horizon because the sinking $CaCO_3$ skeletons dissolve in the undersaturated water before they reach the sea bed. Only 20% of the ocean's $CaCO_3$ enters sediments because 80% of the ocean floor lies below the saturation horizon. *Red clay* (alumino-silicate debris) sinks to the sea bed at a rate of flux of c. $0.3 \, g \, cm^{-2}$ sediment 10^{-3} years and $CaCO_3$ at c. $1.0 \, g \, cm^{-2}$ sediment 10^{-3} years. Above the saturation horizon therefore, sediments contain 3 parts clay to 10 parts $CaCO_3$ and are called *carbonate oozes*. Below it they consist only of red clay (Figure 6.4) because all the $CaCO_3$ has been dissolved before it reaches the sea bed.

Diatom ooze contains the silica frustules of diatoms, and *radiolarian ooze* contains the silica skeletons of radiolarians. These silica-rich sediments occur beneath areas of enhanced upwelling in the equatorial and north Pacific, on the west coasts of South America and Africa, in the Indian Ocean, and as a broad band in the Antarctic. Upwelling also occurs in the Atlantic, but silica-rich sediments are not common there because Atlantic waters have relatively low dissolved silicate levels.

Pelagic sediments cover 74% of the sea bottom, and of these globigerina ooze (46%) and red clay (38%) are the most abundant (Figure 6.3). Diatom ooze occurs as a continuous band around Antarctica and in the North Pacific. Radiolarian ooze occurs mainly in the equatorial Pacific, and pteropod ooze only in the Atlantic.

The boundaries between the sediments are not distinct but there is little overlap. Each ooze has a characteristic range of colour. Globigerina ooze is milky-white, rose, yellow, or brown, and near land a dirty white-grey or blue, while red clay is brick-red in the North Atlantic and chocolate-brown in the South Pacific. Globigerina and pteropod ooze may be sandy (1000– 1 μm diameter) while red clay is very fine (10– < 0.1 μm).

Other less abundant pelagic sediments are coral reef debris from slumping around coral reefs, coral sands, white coral muds, and oolites (Chapter 10).

Manganese nodules

Manganese nodules are solid pebble-shaped objects ranging in size from less than 1 cm to more than 10 cm in diameter, that occur at or near the surface of deep sea sediments particularly in the Pacific Ocean. They are often nearly spherical, but can also be asymmetrical about the sediment-water interface or elongate and sausage-shaped (Moorby, in Cronan, 1980).

It is still not known exactly how manganese nodules form (Cronan, 1980). Their constituent elements (Mn, Ni, Co, Cu, Fe) are present in higher concentrations in sediments and in the nodules themselves than they are in sea water. There are three hypotheses to account for this (Broecker, 1974). The first suggests continued weathering as a source, in which Mn and Fe associate with the lightest particles carried down by rivers which are then spread evenly over the sea bed. The second is called *secondary enrichment*. Aerobic bacteria in sediments deplete oxygen until the deeper layers become anaerobic, whereupon insoluble Mn^{4+} changes to soluble Mn^{2+}. The Mn^{2+} is then carried in solution in pore water to the upper aerobic sediment layers again and reprecipitates as Mn^{4+} leading to manganese enrichment. The third hypothesis suggests that the metals are added to the ocean at the mid-oceanic ridges where new crustal rocks are formed and in hydrothermal vent fluid.

There are other problems. Sedimentation rates of sediments in manganese-rich areas are about 0.3 cm 10^{-3} years but the growth rate of nodules is only 0.3 mm 10^{-3} years. Furthermore, manganese accumulates more rapidly in these sediments than in the nodules themselves. Perhaps Mn, Ni, Co, and Fe, like thorium, rapidly adsorb onto particulate material and enter the sediment in that way. It is not even clear why nodules are usually found at the sediment surface. The manganese nodule story is therefore still something of a mystery.

Sedimentation rates in the deep sea

Sedimentation rates in the deep sea range from $c.$ 0.1 to $c.$ 3.0 cm 10^{-3} years with an average of about 0.5 cm 10^{-3} years for aluminosilicate debris (red clay). This is the rate at which particulate material is added to the sediment from the water column. The rate at which material sinks through the water column is theoretically derived from Stokes' law, which gives the terminal velocity v of a particle as

$$v = \frac{2(d_s - d_f)gr^2}{9\mu}$$

where r = radius of particle, μ = viscosity of sea water, d_s = density of the falling particle, d_f = density of sea water, and g = force due to gravity. However, it is not easy to relate this to rates of settling through the water column, because not enough is known of the size, shape and density of settling particulate material under natural conditions.

There are three sedimentary stratigraphic markers that occur in deep sea sediments which enable one to compare sedimentation and hence rates of incorporation of particulate material into sediments in different parts of the ocean. These are the magnetic reversals that are clearly defined near the mid-oceanic ridges, faunal extinctions and appearances at well-defined sedimentary horizons, and climatic changes—the ice ages for example. Climatic changes are detected by changes in the $CaCO_3$ content of sediment cores, in $^{18}O/^{16}O$ ratios of shells, and by bands of coarser-grained material.

Radioactive dating methods are also available, and are more accurate. Carbon-14 (^{14}C) is formed in sea water from carbon-12 (^{12}C) by bombardment with cosmic rays from space. The $^{14}C/^{12}C$ ratio in the $CaCO_3$ shells in sediments, when compared with the same ratio in sea water, gives accurate dates to 40 000 years before present (b.p.). Two other methods rely on the decay of uranium-238 and uranium-235. ^{238}U and ^{235}U were produced from hydrogen in the centre of stars which then exploded. They became incorporated into our solar system, and are slowly decaying to the stable element lead. However, their half-lives are so long (^{238}U: 4.49×10^9 years; ^{235}U: 7.13×10^8 years) that they are still present on earth in appreciable concentrations. ^{238}U and ^{235}U are very soluble in sea water but their daughter decay products, thorium-230 and protoactinium-231 respectively, are insoluble and so adsorb on to particulate matter. Since the half-lives of ^{230}Th and ^{231}Pa are short (7.5×10^4 and 3.25×10^4 years), their adsorption gives a method for dating sediments. Particulate matter sediments through the water column and enters the sediments. The age of a sediment is then obtained by comparing the $^{238}U/^{230}Th$ ratio in sea water with that in sediments at different depths. The same comparison is also made for the $^{235}U/^{231}Pa$ ratio. The ^{230}Th method is accurate to 400 000 years b.p. and the ^{231}Pa to 150 000 years b.p.

All three methods rely on radioactive atoms (^{14}C, ^{230}Th, ^{231}Pa) being formed at a constant rate in the sea water, but not being formed at all in sediments—only decay taking place. There is good agreement between these three methods when comparisons have been made on the same core. For example, cores from the Caribbean Sea give sedimentation rates of 2 to 3 cm 10^{-3} years, which is a fairly typical value for the deep sea.

Sediment characteristics, shear strength, erosion

The properties of sediments on the sea bed are scientifically and economically very important. They determine whether sediment erosion and transport will take place, control the chemical and physical processes that occur during diagenesis, and limit the abundance and types of benthic animals, plants and microorganisms.

Particle size is one of the most obvious attributes of sediments and is the basis for their classification into mud (clays and silts), sand and gravel (Figure 6.5). The particle size of sediments is measured as the diameter of the particles in mm or in phi (ϕ) units ($\phi = -\log_2 (\text{mm})$), and the following parameters are usually quoted: mean, median, sorting (standard deviation), skewness and kurtosis. The mathematics of these parameters are complicated and are explained in Snedecor and Cochrane (1980), Folk (1980), and Sokal and Rohlf (1981). They can be calculated graphically or algebraically. The *mean* is the average particle size, and the *median* divides

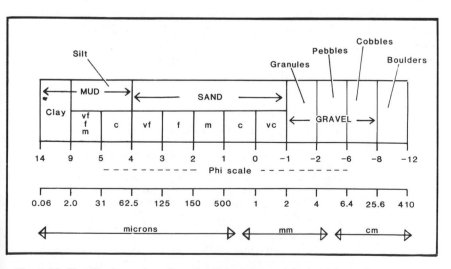

Figure 6.5 Classification and grading of sediments by particle size. *Vf*, very fine; *f*, fine; *m*, medium; *c*, coarse; *vc*, very coarse (after Folk, 1980).

the frequency distribution of the particles into two halves. If the size distribution follows the *normal curve* the median equals the mean. The observed particle size distribution can differ from the shape of the normal curve having the same mean and standard deviation in two ways— *skewness* and *kurtosis.*

Skewness occurs if the size distribution is peaked towards the larger or smaller particle sizes. It can be more peaked towards the larger particle sizes (smaller ϕ) with a tail in the finer particle sizes (larger ϕ). Here the median is less than the mean on the ϕ scale, and the distribution is called *negatively skewed.* A *positively skewed* distribution on the ϕ scale has its peak at the smaller particle sizes (larger ϕ) and tail in the bigger particle sizes (smaller ϕ). On the ϕ scale its median is greater than its mean.

Kurtosis measures the symmetrical flatness or peaking of the observed distribution in the central and peripheral parts of its distribution. An observed distribution is *leptokurtic* if it has a higher central peak falling rapidly on either side of the mean to longer tails, when compared to an equivalent normal curve having the same mean and standard deviation. Conversely an observed particle size distribution is *platykurtic* if it has a lower central peak, is flat topped, and tends to be convex with little or no tails at the extremes of the distribution.

Sediments are often classified into fifteen different *textural groups* based on their proportions of gravel, sand and mud. The groups are often displayed on a triangular diagram in which two sides have a scale of % gravel and one side has a scale representing the ratio of sand to mud. This diagram has been rearranged to form Table 6.1. The textural groups are used in coastal waters to qualitatively describe sediment distribution on the sea bed (Meadows and Tufail, 1986).

Table 6.1 The 15 major textural groups or types into which sediment is classified based on the % gravel and the sand:mud ratio (gravel: > 2 mm; sand: 62.5μm $- 2$ mm; mud: $< 62.5 \mu$m). M = mud, m = muddy; S = sand, s = sandy; G = gravel, g = gravelly, (g) = slightly gravelly. Mud = silt + clay (after Folk, 1980).

Sand:mud Ratio	Percentage gravel				
	< 0.01	$0.01-5\%$	$5-30\%$	$30-80\%$	$> 80\%$
$< 1:9$	M	(g)M	gM	mG	
$1:9-1:1$	sM	(g)sM			
$1:1-9:1$	mS	(g)mS	gmS	msG	G
$> 9:1$	S	(g)S	gS	sG	

The *porosity* (P) of a sediment is the percentage of its total volume that is occupied by interstitial spaces between the particles. $P = (V_v \times 100)/V_t$ where V_v = volume of interstitial spaces and V_t = total volume of sediment sample. The interstitial spaces are partially air-filled in intertidal sands when the tide uncovers the sediment, but are liquid-filled in subtidal sediments unless gas is produced within the sediment by chemical or microbiological means. The water content of a sediment is the weight of the water in the interstitial spaces of a sediment sample as a percentage of dry weight of the sediment sample.

The *permeability* of a sediment is largely determined by particle size—the bigger the particle size, the greater the permeability. This is common sense, since water drains more quickly through sand than through mud. However, biological activity in the form of animal secretions or microbiological growth can clog the interstitial spaces of a sediment and hence reduce its permeability (Meadows and Tufail, 1986). This may lead to anaerobic conditions which will have major effects on animals and microorganisms living in the sediment and on the early stages of sediment diagenesis. *Permeability* (K), sometimes called the *permeability coefficient*, is given by

$$K = \frac{QL}{THA}$$

where Q = the volume of water flowing in time T across a cross section of sediment whose area is A, along a length or height of sediment L. H is the hydraulic head (difference in water height causing water flow). Both H and L are usually taken in the vertical direction. Typical values of the permeability coefficient (K) are 10 to 1000 mm s^{-1} for gravels, 0.01 to 10 mm s^{-1} for sands, 10^{-2} to 10^{-5} mm s^{-1} for silts, and less than 10^{-5} mm s^{-1} for clays (Smith, 1981).

The *shear strength* of a sediment is an estimate of the lateral pressure needed to break a block of sediment or to destroy its structure. It is measured by simple cone or vane penetration devices in the field, or by rather more complicated equipment in the laboratory. Shear strength is measured in units of pressure (force/unit area) as kg cm^{-2} or kN m^{-2} (1 kg cm^{-2} = 98.1 kN m^{-2}). Shear strength is related to particle size, inter-particle binding, porosity, water content and biological activity, although the relationships are not entirely predictable (Lambe and Whitman, 1979; Meadows and Tufail, 1986). Shear strength increases as inter-particle binding increases, and as porosity, water content and particle size decrease. Biological activity in the form of burrows produced by macrobenthic invertebrates and secretory materials produced by invertebrates and

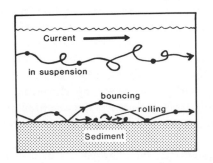 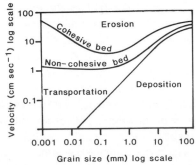

Figure 6.6 *A*: Three ways in which particles are eroded and transported from sediments at and above the critical erosion velocity. *B*: Critical erosion velocities above which erosion and transportation take place and below which deposition takes place, for sediments having different particle sizes. (after Hjulström, 1939; Sundborg, 1956; Meadows and Tufail, 1986).

microorganisms also tends to increase the shear strength of a sediment.

The *critical erosion velocity* is the water velocity in the water mass outside the boundary layer at which particles first begin to be eroded from the sediment surface. The shear stress along the sediment interface at this point is called the *critical shear stress*. Sediment *erosion* and *transport* take place where seabed currents are faster than the *critical erosion velocity*. This often happens near coastlines and in estuaries and harbours, and can lead to major environmental problems. For example shifting sand banks are a regular feature of some estuaries, and erosion under sea-bed pipe lines carrying oil can cause stress and eventual fracture. Sediment erosion and transport also occur in offshore continental shelf and deep sea environments, but to a lesser degree.

Erosion, transport and subsequent *deposition* depend on horizontal water velocity and particle size (Figure 6.6). Sediment particles are eroded and transported by *rolling* or *bouncing* (*saltating*) along the sea bed or at faster water velocities by *suspension* in the water mass. For a given particle size, the faster water velocities erode particles from the sediment surface, intermediate velocities transport particles from place to place, and the slowest velocities allow deposition to take place again.

Sediment diagenesis

Sediment deposited on the sea bed is progressively buried by material from above that has sedimented through the water column, and so is buried deeper and deeper in the sedimentary strata. This process, which is called *diagenesis*, eventually leads to the formation of rock, and occurs largely by *consolidation* and *cementation* of the sediment fabric (Taylor, 1964; Whitten and Brooks, 1978; Folk, 1980). *Consolidation* (compaction) is the packing of individual grains thus eliminating pore space and expelling pore water. It is brought about by the weight of the overlying sediment or rock. *Cementation* is the sticking together of particles by dissolved substances in pore water or by substances dissolved from the rock itself and then reprecipitated.

Most diagenesis occurs near interfaces of two or three of the following phases: air, fresh water, sea water, and sediment. It involves a complex series of chemical and physical processes. Unstable minerals such as aragonite dissolve, pore spaces are filled with precipitated minerals such as calcite, and the double salt calcium magnesium carbonate (dolomite) is formed. Many of these changes are controlled by the Eh and pH of the sediment and pore water and probably also by biological and microbiological activities, particularly during early diagenesis.

Diagenesis is divided into *pre-burial, early burial*, and *late burial* stages. During *pre-burial*, biological and microbiological activity and bioturbation mix and disrupt the surface layers of the sediment to a depth of about 1 m, Eh and pH undergo rapid change, the concentration of clay minerals alters, and authigenic materials such as phosphorite, glauconite and pyrite are produced. In the *early burial* stages, changes in Eh and pH and clay minerals continue, carbonaceous material is oxidised, sulphides are formed, and compaction and cementation begin. Aragonite begins to form calcite, and calcium carbonate begins to form dolomite. In the *later burial* stages, Eh and pH are relatively less important, compaction and cementation are completed, aragonite is completely replaced by calcite, and dolomite formation continues. The point at which one stage merges into the next is often ill-defined and there may be some overlap.

During the preburial and early burial stages of diagenesis there are clearly defined vertical gradients and maxima of O_2, CO_2, HCO_3^-, NO_3^-, NO_2^-, NH_4^+, Mn^{2+}, Fe^{2+} and CH_4 (Figure 3.7, p. 61). These are associated with vertical gradients in the abundance and activity of the major physiological groups of microorganisms in the sediment (Table 6.2) and with changes in sediment Eh and pH (Figure 3.10, p. 65), and are almost certainly very important during the first stages of diagenesis. They start with the aerobic zone at the sediment water interface and pass through the increasingly anaerobic nitrate reduction and sulphate reduction zones to the carbonate reduction zone where methane is generated by methanogenic bacteria (Table 6.2).

Table 6.2 Diagenetic sediment zones, chemical species released by mineralisation of organic matter, and physiological groups of bacteria (modified from Redford, 1958; Stanier, Adelberg and Ingraham, 1977).

Sediment zone (increasing sediment depth)	Chemical species released from organic material by mineralisation	Physiological groups of bacteria
Aerobic zone	$CO_2, NH_3, H_3PO_4, SO_4^{2-}$	Aerobic heterotrophs
Manganese reduction	$HCO_3^-, NH_4^+, HPO_4^{2-}, Mn_4^{2+}$	Aerobic + anaerobic heterotrophs
Nitrate reduction	$CO_2, HCO_3^-, NH_4^+, N_2, HPO_4^{2-}$	Anaerobic denitrification by heterotrophs
Iron reduction	$HCO_3^-, NH_4, HPO_4^{2-}, Fe^{2+}$	Anaerobic sulphate reduction
Sulphate reduction	$CO_2, HCO_3^-, NH_4^+, HPO_4^{2-}, HS^-$	*Desulfovibrio*
Carbonate reduction	$CO_2, HCO_3^-, CH_4, NH_4, HPO_4^{2-}, CH_4$	Anaerobic methanogenesis (methane production) *Methanobacterium*

B. The benthos

Infaunal and epifaunal benthos, size classification

Benthic animals live in or on the sea bed, and since the sea bed is mainly covered by sediments this means that most of them live in or on sediments. Animals living on the sediment surface are called *epibenthic* or *epifaunal*. Animals living in the sediment are called *infaunal* or sometimes *sedimentary*. Sediments from all depths contain epifaunal and infaunal animals, albeit at different densities (Table 6.3). Benthic animals are often divided by size into the megafauna ($> c.$ 20 cm), macrofauna ($c.$ 20 cm–$c.$ 0.5 mm), meiofauna ($c.$ 0.5 mm–$c.$ 50 μm) and microfauna ($c.$ 50 μm–$c.$ 5 μm). There is sometimes disagreement about the exact size divisions, and the microfauna are often placed within the meiofauna.

Table 6.3 Abundance of continental shelf and deep-sea benthic animals (after Heezen and Hollister, 1971; Herring and Clarke, 1971; Sverdrup, Johnson and Fleming, 1942).

Benthic zone	Depth (m)	Mean biomass (grams dry weight animals/m^2 sediment surface)
Continental shelf	0–200 m	200
Continental slope, rise	200–3000 m	20
Abyssal plains	> 3000 m	0.2

Macrofauna, food, bioturbation

Sedimentary macrofauna living in or at the surface of sediments include all the major invertebrate groups and are described mainly from the continental shelf. They feed in one of three ways: by filter feeding, browsing, or ingesting deposited material on or in the sediment. *Filter feeders (suspension feeders)* filter small particles in suspension using fans, sieves or nets (Figure 6.7). Many molluscs (*Mytilus edulis*) polychaetes (*Sabella pavonina*), sponges (*Suberites domuncula*), and ascidians (*Ciona intestinalis*) feed in this way. They all use ciliary action to create a water current across moving strands of mucus that are themselves moved by cilia towards the mouth. The small particles are caught on the mucus and eventually eaten. Many species live in the sediment, only extending their feeding organs into the water, for example the bivalve *Tellina tenuis*. Organic material is therefore taken into the sediment, and either used in body growth or voided as characteristic faecal pellets at the surface or within the sediment. The whole process is remarkably efficient at converting micron-sized particles into animals of cm size in one step.

Browsers are usually active mobile species that move across the sediment

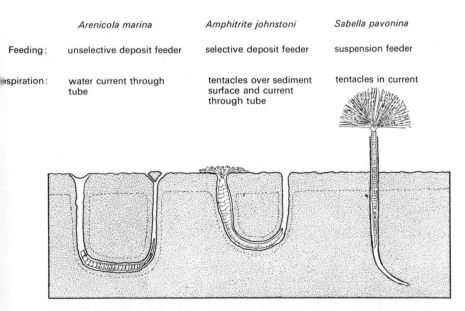

	Arenicola marina	*Amphitrite johnstoni*	*Sabella pavonina*
Feeding:	unselective deposit feeder	selective deposit feeder	suspension feeder
Respiration:	water current through tube	tentacles over sediment surface and current through tube	tentacles in current

Figure 6.7 Feeding and respiration of three common muddy sediment animals. Inshore continental shelf waters, Europe.

E

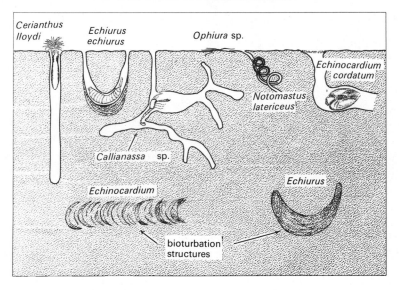

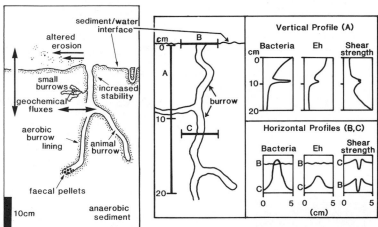

Figure 6.8 *Upper diagram*: Bioturbation by continental shelf burrowing invertebrates and bioturbation structures caused by animals that have subsequently died—Northern European continental shelf species (after Reinecke and Singh, 1973). *Lower left diagram*: Effects of bioturbation on erosion, geochemical fluxes, sediment stability and degree of aerobicity (Meadows, 1986). *Lower right diagram*: Effects of bioturbation on bacterial numbers, and on Eh and shear strength profiles within sediments. Three profiles are shown: *A* is a vertical profile from 0 to 20 cm that crosses a burrow; *B* and *C* are horizontal profiles of 5 cm that cross a burrow at the sediment surface (*B*) and at about 12 cm (*C*).

The changes in bacterial numbers, Eh and shear strength along the three profiles are shown on the right of the diagram (Meadows and Tait, 1985).

surface eating organic material. Many amphipods isopods and gastropods fall into this category.

Deposit-feeding animals eat particles at the sediment surface or within the sediment itself. Again there are representatives from most invertebrate groups: crustaceans (*Corophium volutator*), polychaete annelids (*Arenicola marina*), molluscs (*Scrobicularia plana*), and holothurians (*Holothuria nigra*). Many deposit their faecal pellets at the surface.

Filter feeders are more common in sandy sediments and deposit feeders in finer muds. Sandy sediments usually occur in high energy areas where fine particles are carried into suspension by water currents or waves. This is more suitable for filter feeders. Muddy sediments usually occur in low energy areas where there is less water movement, and so fine particles and detritus settle to the sea bed. This is more suitable for deposit feeders.

Animals and sediments can be regarded as an interacting system. Sediment type influences which animals can live there, and animals alter the physical and chemical structure of the sediment (Meadows, 1986). Animal burrowing and activity in sediments—often called bioturbation—can have major effects (Figure 6.8). Burrow ventilation increases the oxygen content and Eh of the burrow lining, and microbial activity is stimulated locally (Meadows and Tait, 1985; Meadows and Tufail, 1986). The binding materials used by animals in constructing their burrows increase the shear strength of the sediment and reduce erosion. Some larger species of holothurians, polychaetes and crustaceans burrow deep into the sediment, eat anaerobic sediment at depth, and defaecate at the surface. Rhoads (1974) calls these *conveyor-belt* species. The burrowing activity of the holothurian *Molpadia oolitica* increases the spatial heterogeneity at the sediment surface. This species forms a cone of faecal pellets which is surrounded by an annulus of unconsolidated sediment forming a depression. Filter feeding species colonise the cone but not the annulus. All of these biological activities in sediments are now realised as being of potential importance to sediment stability around man-made objects on the sea bed and to the early stages of sediment diagenesis (rock formation) and oil formation (Meadows, 1986; Meadows and Tufail, 1986).

Meiofauna

Meiofaunal animals live in the interstices of mud and sand and are ubiquitous both in freshwater and marine environments (Figure 6.9). Their abundance is often very high (Table 6.4) (Fenchel, 1978; McIntyre, 1969). The meiofauna is a most interesting and diverse group of animals that are

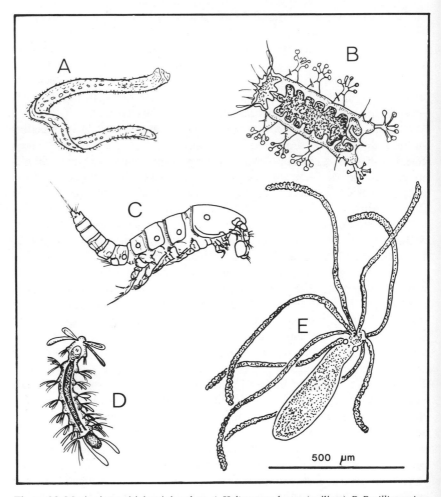

Figure 6.9 Marine interstitial meiobenthos. *A, Helicoprorodon* sp. (a ciliate); *B, Batillipes mirus* (a tardigrade); *C, Tachidius discipes* (a harpacticoid copepod); *D, Nerillidium troglochaetoides* (an archiannelid); *E, Halamohydra* sp. (a hydroid coelenterate) (after Gurney, 1932, Swedmark, 1964, and Fenchel, 1978).

all specialised in one way or another to their unusual habitat (Swedmark, 1964). Meiofaunal animals are either temporary or permanent. The temporary ones are the young of the macrofauna and are sometimes very abundant. In the Wadden sea, Denmark, for example, 72 000 newly metamorphosed *Cerastoderma edule* (the edible cockle) are recorded per m^2

Table 6.4 Meiofaunal abundance (McIntyre, 1969).

Habitats	Number m^{-2}
Intertidal zone	1.1×10^4 to 1.6×10^7
Continental shelf	4.0×10^3 to 3.2×10^6
Abyssal plain	1.0×10^4 to 1.7×10^5

in August but all of them have disappeared by autumn. The permanent meiofauna includes almost all the major metazoan phyla. The most important groups are nematodes, harpacticoid copepods, ostracods, archiannelids, polychaetes, tubellarians and ciliates. Other less common but taxonomically interesting groups are the gastrotrichs, rotifers, tardigrades, kinorhynchs and gnathostomulids.

The main factors determining the abundance, species composition and adaptations of meiofaunal communities are particle size, salinity, degree of sediment anoxia, and habitat (intertidal, continental shelf, deep sea).

The particle size of sediments determines the nature of the meiobenthic community as follows. Interstitial animals living in clean sand whose median particle size is greater than about 100 μm are often long and thin (200 μm–3 mm). Some of the ciliates are even larger! Other adaptations include loss of organs and retention of larval features, external ciliation and gliding (annelids), flat ribbon-shaped bodies (ciliates), attachment organs and claws (turbellarians, gastrotrichs), and adhesive pads (tardigrades).

In sand of median size 100 to 200 μm, ciliates are generally the most important group followed by harpacticoids, turbellarians and gastrotrichs. Above a median particle size of 200 μm ciliates are still common but there is also a rich fauna of metazoans (archiannelids, gastotrichs, harpacticoids, nematodes, oligochaetes, ostracods, tardigrades and turbellarians). Very coarse sands contain a number of unique species.

The meiobenthos is very different in muddy or clay sediments having a median particle size of less than 100 μm and is dominated by burrowing nematodes and harpacticoids.

The main factors controlling distribution on intertidal beaches are probably temperature, salinity, oxygenation and sediment particle size. Some species have unusually wide tolerances. The harpacticoid genus *Platychelipus* can withstand freezing sea ice at − 9 °C for 9 h (Barnett, 1968).

Experimental studies on *Protodrilus* species show that they prefer high oxygen tensions and certain types of bacteria, and also have clearly defined temperature preferences (Gray, 1974, for references).

On subtidal continental shelf sediments the same factors are important, but obviously the range of variation in temperature and salinity are much less. Little is known about the environmental variables that control meiofaunal distribution in the deep sea.

The meiofauna shows marked seasonal peaks of abundance in intertidal and continental shelf sediments. There are usually one or two peaks in spring, summer or autumn. Nothing is known of the deep sea. The peaks represent successive generations, and breeding may occur two or three times throughout the year.

Meiofaunal animals are preyed upon by a number of fish including flat fish and gobies, by hydroids and by polychaetes (*Nereis diversicolor*). Some meiofaunal nematodes, turbellarians and tardigrades are themselves predators of other members of the meiofauna. The nematode *Halichoanolaimus* eats other nematodes and tardigrades eat nematodes and rotifers. The best known example is probably the widely distributed brackish-water interstitial coelenterate *Protohydra leuckarti* (Maas, 1966). This species feeds voraciously on harpacticoids and nematodes and may limit their abundance in spring when it reaches densities of 2×10^5 individuals per m^2.

The vertical distribution of the meiofauna in sediments is very localised. They usually occur within a few cm of the sediment surface whether intertidally or in the deep sea, and there are also species differences. For example, 50–70% of meiofaunal foraminiferans in sediment from 1300 m in

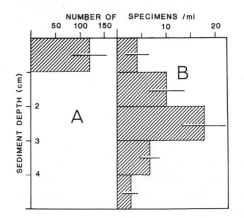
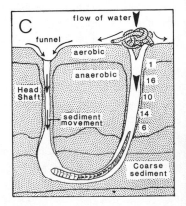

Figure 6.10 Vertical distribution of the meiofauna in sediments. *A* and *B*: deep-sea sediment, 1300 m, Porcupine Seabight, North East Atlantic. Foraminifera: *A*, *Ovammina* sp., *B*, *Pseudononion* aff. *japonica* (after Gooday, 1986). *C*: Intertidal sediment, Island of Sylt, North Sea. Localisation of the gnathostomulid *Haplognathia rosea* in the tail shaft of *Arenicola marina* (no./20 ml sediment) (after Reise, 1985).

the Porcupine Seabight (North-East Atlantic) occur in the top 1 cm, although some species have peaks below the surface (Figure 6.10) (Gooday, 1986). The meiofauna is also very abundant where macro-invertebrates burrow (Reise, 1985). There are many more nematodes where bioturbation by *Arenicola marina, Corophium volutator* and *Pygospio elegans* is heavy, and some meiofaunal species are localised at specific points in *Arenicola* burrows (Figure 6.10).

Macro-algae, communities, production, production/biomass ratio

The only macro-benthic plants below ELWS on the continental shelves are the macro-algae (kelps), a few marine grasses, and the calcareous red algae that encrust pebbles and stones within about 10 to 20 m of the surface. The kelps grow from low water to a depth of about 2 m in turbid water, but to a depth of more than 30 m in clear water on exposed coasts. They are most abundant in temperate climates and grow attached to rocks or large stones. They are an important food for invertebrates (*Echinus esculentus*), and their fronds and holdfasts are well defined habitats for epiphytic plants (other algae) and animals (small crustaceans, polychaetes, molluscs, and Polyzoa).

A forest of sub-littoral kelps often develops with a well-defined canopy that can reduce light intensity to half its surface value. In west-coast Scottish waters *Laminaria cloustoni* and *L. saccharina* (and to a lesser extent *Saccorhiza bulbosa*) may grow to a length of 3 m, and be so dense that a diver has to cut his way through them. In South Africa and in the Pacific, forests of the giant kelps *Macrocystis pyrifera, Nereocystis luetkeana* and *Alaria fistulosa* are common. They often grow to 30 m, and one of 300 m is recorded. Growth rates of $25 \, cm^{-1}$ day are not unusual.

Benthic macro-algae have very high rates of production compared with phytoplankton (Chapter 4). They cover about 0.1% of the world's ocean surface, but their production is about 10% of that of the ocean's phytoplankton. In California giant kelp can produce $12 \, kg \, m^{-2} \, yr^{-1}$, which is ten times the annual production of temperate water phytoplankton, although this is an exceptional figure. The explanation probably lies in the constant changing of the water surrounding the kelp by currents and tides, so providing a continuous supply of nutrients.

The potential annual production of macroalgae available for commercial harvesting is high, particularly around North and South America, Iceland, Europe, Scandinavia and the Far East, and largely consists of subtidal kelps (Laminariales) and intertidal Fucoids (Fucales) (Figure 6.11).

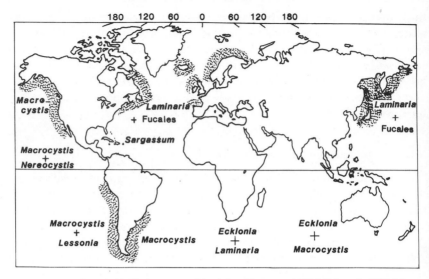

Figure 6.11 Major subtidal kelp communities (after Chapman, 1970; Michanek, 1975; Mann, 1982).

A few marine grasses grow on and stabilise muddy sediments just below ELWS. In temperate waters beds of eel grass *Zostera marina*, and in the tropics beds of turtle grass *Thalassia testudinum* and of manatee grass *Cymodocea manatorum*, are important habitats and food for a range of animals.

Three major subtidal kelp communities are recognised on a global scale. These are the *Ecklonia, Laminaria*, and *Macrocystis* communities (Figure 6.11). *Laminaria* species are dominant on the East and West coasts of the Atlantic, in China and Japan, and also on South African coasts. *Ecklonia* species are common in Australia and South Africa. *Macrocystis*, the giant kelp, forms dense belts on the Pacific coasts of North and South America, around the Falkland Islands and on the east coast of South America.

The productivity of subtidal seaweeds can be very high (Maan, 1982). The *Laminaria* ecosystem in St. Margaret's Bay, Nova Scotia (10 km × 14 km) has a biomass of *c*. 1.5 tonnes m^{-1} of shoreline, of which 84% is *Laminaria* or *Agarum*. In the seaweed zone *Laminaria* productivity is *c*. 1750 g C m^{-2} yr^{-1}. Seaweed productivity over the whole bay is *c*. 600 g C m^{-2} yr^{-1} compared with *c*. 200 g C m^{-2} yr^{-1} for phytoplankton. In other words, in this close inshore environment about 75% of the total primary production is macroalgal. The seaweed itself grows throughout the year, and new tissue is formed at the base and erodes from the tips. *Laminaria* blades are replaced between two and five times a year and so algal detritus, which is readily degraded by fungi and bacteria, must provide an important and continuous input into detrital food webs in the area.

Macrocystis develops flotation organs so that a canopy forms in the top metre of the sea where *Macrocystis* is abundant. Photosynthesis is highest within the top metre, falls to 4 to 20% at 4 m and to 0.5 to 2% at 8 m. Total plant biomass (standing crop) is about $4 \, kg \, m^{-2}$ which is equivalent to about $1 \, kg \, C \, m^{-2}$. Net production can reach 6% of total biomass per day and average production is $c.$ 350 to 1500 $g \, C \, m^{-2} \, yr^{-1}$. On these figures, the production/biomass (P/B) ratio for *Macrocystis* is 0.35 to $1.5 \, yr^{-1}$ while that of *Laminaria* is 2 to $7 \, yr^{-1}$, although *Macrocystis* is often considered to be the more productive species.

Less is known of production by *Ecklonia* beds, although biomass of $7 \, kg \, m^{-2}$ net weight and productivity of $c.$ $1000 \, g \, C \, m^{-2} \, yr^{-1}$ are recorded.

There has been disagreement about additional production from kelps as dissolved organic material (DOM) (Mann, 1982). In *Laminaria*, about 67% is passed to heterotrophs as particulate detrital material from blade breakdown and 33% as DOM. DOM may also be released from cells during photosynthesis.

Sea urchins, otters, and seaweeds

Sea urchins are usually found on subtidal beds of seaweeds where they may feed on detrital material or more usually on the seaweeds themselves (Lawrence, 1975). Sea urchins can control the community structure of the seaweed beds. When *Strongylocentrotus lividus* was removed from subtidal rocks at Friday Harbour, Washington, Pacific coast, USA, there was an initial rapid increase in species diversity followed by domination by two *Laminaria* species. In control areas where the sea urchin occurred at $c.$ 6 animals m^{-2}, the only macroscopic algae were *Lithothamnion* and *Ulva* or *Monostroma* (Paine and Vadas, 1969).

In some areas an inverse relationship between subtidal macroalgae and sea urchins may develop over a number of years caused by increased sea urchin densities and grazing. This relationship may be a long term trend, or develop into an inverse cyclical relationship. Other predators may also affect these changing balances. The sea otter *Enhydra lutris* used to be very abundant on North Pacific coasts from northern Japan to California, but now only occurs in isolated populations. The species regularly eats sea urchins and is one of their major predators. Its effect can be seen by comparing two Aleutian Islands, Amchitka and Shemya. The former is inhabited by otters, while the latter is not (Estes and Palmisano, 1974). At Amchitka, the subtidal cover of kelp is dense and there are few sea urchins. At Shemya, in contrast, there is very little kelp and the sea urchins are extremely abundant, sometimes carpeting the sea bed near the intertidal zone.

Continental shelf benthic communities

Benthic invertebrates on the continental shelf occur in characteristic groups of species that are related to water depth and particularly to sediment type (Figure 6.12). One community may merge into another as the water depth or sediment changes, and intermediate communities are recognised. Thorson (in Hedgepeth, 1957) has distinguished about 14 communities or groups which have a world-wide distribution. The communities are made up of burrowing forms or slow-moving surface dwellers, such as bivalve molluscs (*Mya, Tellina, Cardium, Venus*), polychaetes (*Arenicola, Aphrodite, Nephthys*), echinoderms (*Echinocardium, Spatangus, Amphiura*) and Crustacea (*Corophium, Bathyporeia*). *Tellina* communities, for example, live in clean sand from the intertidal zone to about 10 m, and are recorded from Europe, North America, the Mediterranean, and New Zealand, while *Amphiura* communities live in soft sandy mud, clay, or silt sediments, from 15 m to 100 m, and are recorded from Europe, the Mediterranean, Japan, and New Zealand. Communities are named by the dominant genus: *Macoma* (Bivalvia, Mollusca), *Tellina* (Bivalvia, Mollusca), *Amphiura* (Ophiuroidea, Echinodermata). When the same communities occur in different geographical areas, they contain the same genera but different species (Figure 6.12). Many species disturb the sediment as they move through it or construct permanent burrows (Figure 6.8, p. 120) and the disturbance or bioturbation can sometimes be detected in fossil sediments. The species in some communities grow slowly and are long lived (*Macoma* and *Tellina* communities) while those in others grow quickly and are short lived (*Syndosmya* communities); the population densities of the former remain fairly constant, while those of the latter fluctuate widely. Some communities are an excellent source of food for commercially important fish (*Syndosmya* for flounders and *Amphiura* for haddock). In general, species belonging to Arctic and Antarctic coastal communities have a low productivity, grow slowly, mature late, and have long life cycles. In temperate and tropical coastal waters, some communities have a low

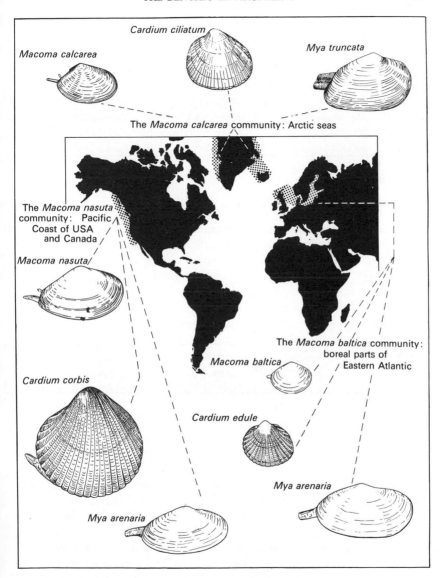

Figure 6.12 Parallel communities of *Macoma* (Bivalvia, Mollusca) on continental shelves: *Macoma nasuta* community on the Pacific coasts of the USA and Canada, *Macoma calcarea* community in Arctic seas, *Macoma baltica* community in the boreal (northern) parts of the Eastern Atlantic.

productivity (*Macoma* and *Tellina*), while others are highly productive (*Syndosmya* and tropical crab communities).

Highly mobile species, such as many benthic decapod Crustacea (*Mya squinado, Cancer pagurus*) do not come into Thorson's classification, as they undertake extensive seasonal migrations on and off shore. For example, around Britain planktonic larvae of *C. pagurus* hatch in summer from eggs carried by the female while she is inshore (0–20 m); the adults then moult, migrate offshore, and remain there during the winter; the following spring they migrate inshore again. Tagged adults have migrated 100–200 km in the North Sea.

Larvae of benthic invertebrates

Many bottom-living invertebrates produce larvae that are planktonic. The larvae usually live near the sea surface, are photopositive and geonegative, and form the major part of the meroplankton.

They may be carried many kilometres by tides and currents, in some instances even across the Atlantic (Scheltema, 1971). They also have limited powers of locomotion, and can select their appropriate habitat within a circumscribed water mass. As they approach metamorphosis, they become photonegative, swim towards the bottom, and periodically alight to select the most suitable sediment in or on which to settle and metamorphose (Meadows and Campbell, 1972*a*, *b*). After metamorphosis, the adults remain within one local area, but even then the adults of some species retain considerable powers of locomotion.

In the Arctic and Antarctic, and also in the deep sea, the larvae of benthic invertebrates are usually liberated from the parent at a late stage in development, and they have a short or non-existent planktonic larval stage varying from minutes to hours. As one moves into boreal (cold temperate) waters and then into the tropics, many more species produce larvae that have a long planktonic life (weeks to months). However there are a number of exceptions; for instance some tropical tunicates produce larvae that metamorphose within a few hours.

The larvae of some species such as cirripedes (barnacles) are gregarious and settle near adults of their own species. Extracts of adult barnacles contain a protein, arthropodin, which is extremely attractive to their cyprid larvae as they settle and is the chemical basis of gregariousness in the species (Crisp and Meadows, 1962, 1963). Similar gregarious responses probably exist in other groups such as the polychaetes, molluscs and echinoderms.

Diversity and stability in benthic communities

Community diversity depends on the number of species and individuals in the community at a given point in time, and is mathematically well documented. A community with a high diversity has many species and relatively few individuals per species. A community with a low diversity has fewer species but more individuals per species. There are many diversity indices that quantify community diversity in marine, fresh water and terrestrial environments (Pielou, 1977; Levinton, 1982). The most frequently used ones are Simpson's index (Simpson, 1949) and several that are based on the Shannon function (Shannon and Weaver, 1949). The Shannon function is given by

$$H = -\sum \frac{n_i}{N} \log \frac{n_i}{N}$$

where n_i = the number of individuals in the ith species and N = the total number of individuals in all the species. Commonly chosen logarithm bases are 2, e and 10 (Pielou, 1977).

Simpson's index is

$$C = \sum \frac{n_i(n_i - 1)}{N(N - 1)}$$

The Shannon function and Simpson's index are closely related mathematically (Pielou, 1977).

Community stability is a time-dependent concept, and is less well defined. A community is stable if its diversity, species composition, or numbers have changed little over a given time. Community stability should not be confused with environmental stability. An environment is stable if it changes little over a given time and unstable if it changes significantly.

In general, community diversity is high in the tropics, intermediate in temperate waters and low in the arctic and antarctic (Figure 6.13). Pacific Ocean communities are usually more diverse than Atlantic ones and deep-sea communities more diverse than inshore or estuarine ones.

Estuarine benthic communities have a very low diversity but are fairly stable although living in a very unstable environment. This is probably because their constituent species reproduce quickly. Deep sea communities have high diversity indices, live in a very stable environment and reproduce very slowly. Shallow-water temperate communities have an intermediate diversity and stability. Coral reefs and to a lesser extent mangrove swamps

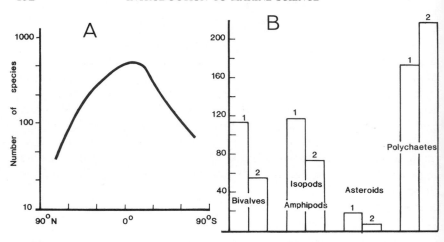

Figure 6.13 Community diversity. *A*: Number of species of bivalve molluscs at different latitudes (Modified from Stehli, McAlester and Helsley, 1967, and Levinton, 1982). *B*: Numbers of species at (1) Friday Harbor, Washington (Pacific Coast) and (2) Woods Hole, Massachusetts (Atlantic Coast) of North America (after Levinton, 1982).

have a high diversity, but their stability and that of their environment is low. The problem is clearly complex, and there are different views about the underlying causes.

High community diversity may lead to high community stability, because each predator has a wide choice of prey species, each prey species is regulated by several predator species, and the elimination of one predator or prey species will make little difference. However, complex ecosystems (coral reefs) may themselves have a physical complexity of structure which provides a wide range of habitats for a large number of species (Levinton, 1982). Community diversity may therefore reflect the heterogeneity of the environment.

Sanders (1968) has suggested that environments which are stable for long periods of time such as deep-sea sediments may allow very diverse communities to develop. This is because the long-term environmental stability exerts little ecological stress on the community. Sanders calls this a *stability–time* hypothesis. He suggests that in the constant conditions on the deep-sea floor, stresses between species such as competition and fluctuating predator/prey relationships have become reduced, and more and more species enter the community. However, the species are now tolerant of only a narrow range of conditions.

Valentine (1973) suggests that in the tropics where primary production is the same throughout the year (*trophic constancy*), benthic species evolve specialised food preferences. This allows complex benthic food webs with efficient energy utilisation to develop, which in turn leads to highly diverse communities. In the Arctic where primary production is high for a short time each year, fewer benthic species occur but these species have more generalised feeding habits, and benthic communities in these environments have a low diversity.

There is only a small amount of evidence available on the *temporal stability* of benthic communities and the concept has not been adequately formalised. There are two well-known examples, one involving decades and one hundreds of millions of years. The eel grass *Zostera marina* occurs in shallow water (< 10 m) beds in sediments and used to be widely distributed on the Atlantic and Pacific coasts of North America. The eel grass community is a diverse one and the subtidal grass beds provide local environments that are suitable for a number of benthic species. *Zostera* suffered a catastrophic decline in the 1930s and since then has slowly recovered. This had major effects at the time. Community diversity was dramatically reduced in and near the eel grass communities and some eel grass communities disappeared completely. The local benthic environment changed dramatically, and sediment erosion increased. There is also evidence on a geological time scale. Over the last 600 million years there have been long-term trends in the average number of benthic species in marine environments (Bambach, 1977). The mean number of species in offshore deep water communities has increased from about 20 to 60, in the more variable near-shore environments has increased from about 13 to 40, and in the high-stress inshore environments has remained approximately the same at about 7 to 10. These short- and long-term changes emphasise the importance of long term environmental stability in maintaining community stability and thus allowing present-day community diversity to develop.

Ecological effects of increasing water depth

As one moves from the continental shelf down the continental slope to the abyssal plains, a number of ecological changes occur, although many of them are poorly understood.

Firstly, the transfer of organic material from the surface waters to the sea bed and back again is reduced as water depth increases. This two-way exchange is sometimes called benthopelagic or benthic-pelagic coupling. In addition, transfer in shallow waters will be low during summer, when the thermocline prevents vertical mixing, and high during winter when the thermocline does not exist. The effect of increasing depth is shown by the following example. The proportion of organic matter derived from dead plankton and sinking to the sea bed decreases with increasing water depth. 20% to 50% of the primary production reaches the sea bed in temperate estuaries and coastal waters, but at below 2000 m this falls to less than 10%. In the shallow water of Long Island Sound ($<$ 100 m) surface primary production is c. 70 g C m^{-2} yr^{-1} of which c. 25 g C m^{-2} yr^{-1} (36%) reaches the sea bed. In the Sargasso Sea, at the centre of the North Atlantic ($>$ 2000 m) the values are c. 130 g C m^{-2} yr^{-1} and c. 4 g C m^{-2} yr^{-1} (3%). In addition, at greater depths the organic matter is usually more refractory (not easily broken down by organisms) because most of the non-refractory material has been metabolised by heterotrophic organisms and bacteria in the water column.

Sinking rates of detrital material range from weeks to over a year per 1000 m, although there are fast by-pass routes. Zooplankton faecal pellets fall quickly, and turbidity currents carry material to the sea bed with great rapidity. Recent work at the Institute of Oceanographic Sciences in Britain has shown that in the North East Atlantic, 'fluff' containing high levels of chlorophyll reaches the bottom within days. Fluff levels on the sea-bed peak at the same time as spring phytoplankton bloom. The fluff quickly disappears from the sea bed, and is thought to be eaten by benthic deposit feeders.

The metabolism of benthic communities, measured by oxygen consumption, decreases with increasing water depth. It is 4 to 40 ml O_2 h^{-1} at 10 to 200 m and 0.5 to 4.5 ml O_2 h^{-1} at 1300 to 2900 m. Most of this oxygen is consumed by heterotrophic bacteria, less by the meiofauna and very little by the macrofauna.

The decreasing organic input from the water column reduces the amount of organic material in sediments and this increases the sediment redox state. The organic content of sediments shallower than 100 m are about 4 to 5% and Eh (redox) falls from about $+$ 200 mV at the sediment surface to $-$ 100 mV at 10 cm depth. Below 2000 m, organic content is about 1%, and Eh only drops from $+$ 200 mV to $+$ 50 mV (Meadows and Tait, 1985). The relationship between decreased organic content and higher Eh is probably caused by microbial action. In shallow waters large numbers of aerobic bacteria breakdown the greater amount of organic material using oxygen in the process. Hence Eh, which is closely linked to oxygen content, falls. In deeper-water sediments with a low organic content, fewer aerobic heterotrophic bacteria breakdown less organic material thus using less oxygen. The Eh therefore does not fall so much.

Numbers of species increase with water depth to about 2000 to 3000 m, and then decrease again, although there is considerable variation (Rex, 1981). Invertebrate and fish megafauna as a whole show a slight trend, while the effect is more obvious within taxonomic groups—in the gastropods and polychaetes for example (Figure 6.14). The species diversity of the meiofauna, however, continues to increase with depth. Amongst the macrofauna, the shelf/slope transition at 100 to 300 m marks an important taxonomic division. Continental shelf species of bivalves, crustaceans and polychaetes dominate above this depth, and deep-sea species below it. All of this, of course, should be seen against the dramatic reduction in animal numbers and biomass with depth (Table 6.3). There is discussion about the causes of these effects.

There are interesting similarities and differences between the reproduc-

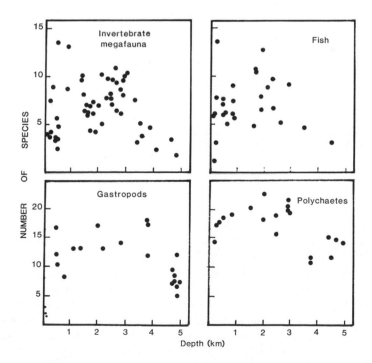

Figure 6.14 Relationships between numbers of species and water depth (after Rex, 1981).

tive strategies of continental shelf and deep-sea benthic invertebrates (Gage, 1985; Grant, 1985; Tyler, Muirhead and Colman, 1985). Continental-shelf species almost always show marked seasonality in their reproductive cycles, producing eggs and larvae during spring and also sometimes in late summer. This is clearly related to environmental clues such as temperature and light. However the deep sea is totally dark, has a constant low temperature of about 2–6 °C, and is regarded as being a very stable unchanging environment. One would therefore expect that benthic species would show no seasonality in their reproductive cycles. This is certainly true of many species of echinoderms (*Benthopecten armatus*), gastropods (*Colus jeffreysianus*) and decapods (*Parapagurus pilosimanus*). These species either produce eggs continuously or produce discrete cohorts of eggs but there is no seasonality in their production (Giese and Pearse, 1974). However, analysis of the reproductive biology of echinoderms from a time series in the NE Atlantic has shown a range of reproductive strategies from the usual continuous reproduction and direct development, to a most unexpected seasonal production of planktonic larvae (Tyler *et al.*, 1982). The reasons for the remarkable seasonality shown by some of these species are not clear at the moment, but the seasonality implies that information of some sort must reach the deep sea on a regular annual basis, and be recognised and acted upon by the benthic fauna living there.

Deep-sea benthic animals

Benthic animals occur in the deepest part of the oceans—the bottom of the oceanic trenches. However, they are much less common in deep sea sediments than on the continental shelf, and of course there are no deep sea benthic algae because there is no light. Photographs of the deep sea floor show that animals, and their tracks, are not as abundant as in shallower waters but are very characteristic (Figure 6.15). Biomass of the larger benthic animals is roughly 200 g dry weight/m² sediment surface on the continental shelf but only 0.2 g m⁻² below 3000 m (Table 6.3, p. 118). In abyssal regions more animals are found beneath high productivity waters of the Antarctic and Arctic than beneath the less productive temperate waters (Table 6.5).

Bathybenthic and abyssobenthic animals belong to most of the major invertebrate groups (Porifera, Coelenterata, Echinodermata, Mollusca, Annelida, Arthropoda) as well as to the major groups of fish (Figure 6.16). Echinoderms are the commonest group, for example ophiuroids (brittle stars) and elasipod holothurians are relatively abundant and eat ooze.

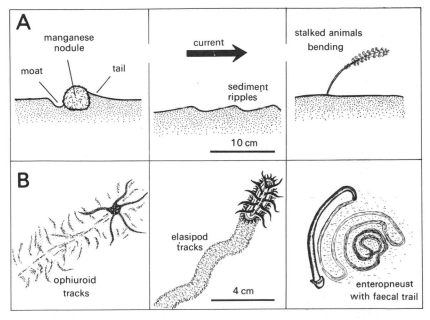

Figure 6.15 Microtopography of, and animals on, the bed of the deep sea (4000–6000 m). *A*. Effect of bottom currents (the stalked organism at top right is probably a pennatulid coelenterate). *B*. Three common benthic invertebrates showing mud tracks and a faecal trail (modified from Heezen and Hollister, 1971).

Table 6.5 Relationship between deep-sea benthos and primary productivity of surface waters (after Heezen and Hollister, 1971; Herring and Clarke, 1971).

Productivity of surface waters	Oceanic area	Benthic animals seen in photographs below 3000 m (number/10 m² sediment surface)
high	Antarctic Arctic of Chile, Peru (upwelling)	4–10
low	South Atlantic South Indian South Pacific	0–3

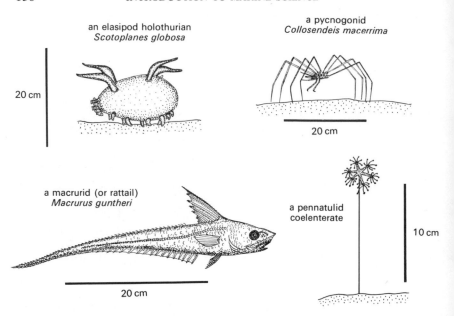

Figure 6.16 Deep-sea benthic animals (modified from Herring and Clark, 1971, and Heezen and Hollister, 1971).

Some species live on the sediment surface, and some burrow into it. Many species are red, grey or black, and sessile forms are often supported above the sediment on long stalks (tunicates (*Styela, Culeolus*), pennatulid coelenterates). Crustaceans and fish are often blind, and some of the latter have long fins on which they stand. Deep sea hexactinellid sponges filter-feed and have a brush-like root in the sediment. Many crabs, prawns and shrimps have long spidery legs, for example those of the pycnogonid *Colossendeis macerrima* extend over 25 cm of sediment. In general, as depth increases, deposit and suspension feeding species increase in abundance, and scavengers and carnivores decrease. Below 5000 m deposit feeders like the elasipod holothurians are dominant. Physiologically, some of these animals are particularly interesting, as their metabolic rate is unexpectedly low; they appear to live longer than shallow water forms, and they are relatively small in size.

Deep-sea cod (Moridae), rat tails (Macruridae) and brotulid fish live on or near the deep sea bottom, but are most abundant on the upper continental slopes in the tropics and sub-tropics. They root in ooze, and eat burrowing invertebrates and surface-dwelling ophiuroids and worms.

Their swim bladders are sound-producing organs that may produce sexual signals.

Adaptations to deep sea life

Buoyancy in deep sea fish without swim bladders is achieved by the presence of more fat and less protein, and by a poorly ossified and lightly developed skeleton in the bony fish, and in the elasmobranchs by high concentrations of the light oil squalene (specific gravity 0.8) in the liver. Many species have bioluminescent organs arranged along their bodies in species-specific patterns, and these may serve as sexual recognition or as lures for prey.

Reproduction at great depths is likely to be more difficult than in shallower waters because of the rarity of animals and the total darkness. Amongst the deep sea invertebrates, one adaptation that is common is for species to produce large yolky eggs which then give rise to advanced young. In deep sea fish, the occurrence of hermaphroditism, dwarf parasitic males, and species-specific patterns of luminescence on the body, may assist in sexual recognition and reproduction. Some species are viviparous (Brotulids), while others produce pelagic larvae that live in the upper 200 metres of the water column (Halosaurs, rat tails). These larvae are exposed to considerable temperature change during metamorphosis to the bottom-living adult stage. In the tropical North Atlantic, for example, the larvae of deep sea angler fish live near the surface at temperatures of 20 to 25 °C, while after metamorphosis the adults live at about 2000 metres depth where the temperatures are about 3 or 4 °C.

The food of deep-sea carnivorous fish is relatively scarce, and so diverse feeding adaptations have been evolved. Many fish are small, being only a few centimetres long, but they have large teeth, flexible jaws with a large gape, and a distensible stomach. The angler fish *Melanocoetus johnsoni*, for instance, can catch and swallow prey twice its size. Luminescent lures are also used by some species to attract prey.

There has been some speculation about the way in which food is transported from the upper layers to the abyssal depths. Clearly there is a constant slow rain of decaying organic material, as witnessed by the pelagic sediments made up of the skeletons of planktonic organisms (foraminiferan ooze, pteropod ooze). However, a series of overlapping vertical migrations by different species has been postulated, in which one species forms the food for the next; in this way food might be transported more rapidly into deeper waters.

Although the design of sampling gear for use at great depths is still in its infancy, many deep sea benthic invertebrates can now be caught by suitable dredges and grabs. Equipment is often designed to remain shut unless actually operating on the bottom so that only deep sea benthic animals are caught (the Sanders epibenthic sledge), and to dig into the sediment to a predetermined depth (the modified anchor dredge operated by the SMBA Dunstaffnage Laboratory, Scotland, digs to 10 cm). Recent advances in the design of this kind of equipment have greatly increased our knowledge of the deep sea benthos. For example, when the Sanders epibenthic sledge was first used off Bermuda, at least one-third of the species caught were new to science, and it is now appreciated that although the deep-sea benthic fauna is very sparse it contains a wide range of species. Other new pieces of equipment now routinely used for obtaining undisturbed samples of deep-sea sediment are the Barnett Watson multicorer (SMBA) and the Spade Box Corer. Remote controlled landers are also used for studying the metabolic activity of deep-sea sedimentary bacteria.

Hydrothermal vents, vent fauna

Hydrothermal vents are holes in the deep-sea floor from which hot water (*hydrothermal fluid*) gushes (Jannasch, 1985; Grassle, 1986). The holes are 20 to 50 cm in diameter and are often surrounded by a chimney which is 1–2 m high. There are very high densities of macrobenthic animals around the vents within tens of metres, in contrast to most areas of the deep-sea floor. This indicates a localised food source which has proved to be based on chemoautotrophic microorganisms most of which are hydrogen sulphide oxidising bacteria. Some of these bacteria may be able to grow at up to 300 °C in the vents. The whole phenomenon is quite remarkable (Figure 6.17).

Many of the animals and bacteria are new to science. *Riftia pachyptila* is a vestimentiferan pogonophore which is up to 3 m long and 5 cm diameter. *Calyptogena magnifica* is a 30 cm-long bivalve mollusc. A new genus of bivalves—*Bathymodiolus* has also been described. The numbers of macrobenthic animals are usually higher around vents which have water high in HS^-, as long as aerobic chemosynthesis can occur.

Hydrothermal vents are of two types—*warm vents* and *hot vents* (Figure 6.17). As sea water penetrates into the crust, it is heated to 350–400 °C, reacts with basaltic rocks and leaches various chemical species into solution (Figure 6.17). The highly reduced hydrothermal fluid rises and reaches the sea floor either directly (hot vents) or after mixing with cold and

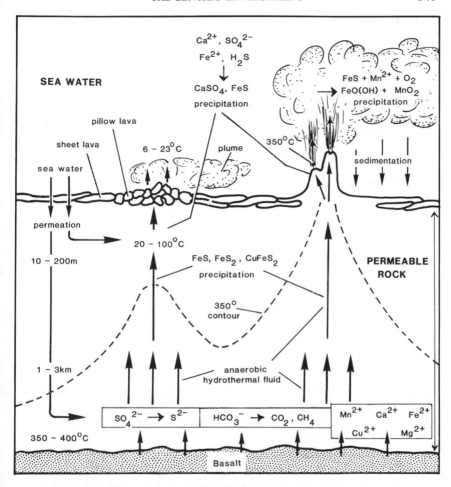

Figure 6.17 Hydrothermal vents at areas of the deep-sea bed where new sea floor is forming (spreading centres). Scheme of inorganic processes taking place and below the deep-sea bed. A warm vent (left, 6–23 °C) and a hot vent (right, ≤ 350 °C) are shown (after Jannasch, 1985).

oxygenated water before emission (warm vents). On mixing, polymetal sulphides and calcium sulphate (anhydrite) precipitate, either within subsurface lava conduits, or as chimneys and suspended particulate matter of the hot vents (Jannasch, 1985). At *warm vents* water is discharged at 6 to 23 °C and flows out of the vent at about 0.5 to 1.5 cm s^{-1}. This produces slowly moving HS$^-$ plumes that contain *c.* 0.05 to *c.* 0.25 mM HS$^-$. The plumes spread over large areas locally, and *Riftia* forests such as those at the

Galapagos Rift vent site are associated with them. Hydrothermal fluid gushes from *hot vents* or *black smokers* more quickly, at 1 to $2\,m.\,s^{-1}$, has higher temperatures (up to $350\,°C$) and higher sulphide concentrations (0.5–10 mM). It does not appear to have mixed with cold oxygenated sea water below the sea bed. As it gushes out, it meets cold sea water and there is a rapid and heavy precipitation of dark grey polymetalsulphide particles which produces the characteristic smoke plume. A chimney progressively forms around the vent by instant precipitation of calcium sulphate combined with polymetalsulphides. The conduits that feed black smokers below the sea bed progressively silt up with this precipitation which leads to the smokers' short life—estimated at 10 to 100 years. Black smokers are often surrounded by warm vents which produce areas of *Riftia* and *Calyptogena* nearby.

Warm and hot vents often occur in aggregates or fields. They have been described off the west coast of Central and North America, at the Galapagos Rift, and at the spreading centres of the East Pacific Rise (1600–3000 m deep). They also occur off Florida in the Caribbean Sea and probably along the Mid-Atlantic ridge and in the Marianas Trench region.

Cold seeps of water from the sea bed at non-hydrothermal vents also have large numbers of macrobenthic organisms near them, but are less spectacular. The energy source here is probably microbial S and SO_4^- reduction, and oxidation of organic matter.

The chemosynthetic bacteria that oxidise HS^- at the hydrothermal vents are particularly interesting. Many are thermophylic with doubling times of 30 minutes, and grow at 80 to $100\,°C$ and possibly at $300\,°C$. They are obligate chemolithotrophic sulphur oxidisers, mainly *Thiomicrospira*, but also *Thiobacillus* and *Hyphomonas*. Aerobic chemosynthesis by these microorganisms occurs in underground chambers and conduits beneath the vents, and also in microbial mats that develop on exposed surfaces near vents. They are also found in symbiosis with invertebrates. However, no estimates of anaerobic chemosynthesis are available at present. The symbiotic chemosynthesis, which is analogous to the symbiotic photosynthesis of zooxanthellae in corals, probably accounts for much of the macrobenthic biomass near vents. The bacteria live inside modified cells of *Calyptogena magnifica* and *Riftia pachyptila*, called bacteriocytes. *Riftia* has no mouth, gut or anus and relies solely on these bacteria for its nutrition. The bacteriocytes are localised in red *trophosome* tissue which is well supplied with blood vessels and may form more than half the mass of the animal. The bivalve *Calyptogena* is less highly specialised, having a mouth, gut and anus, and the bacteriocytes are in the gills.

Three other vent animals have been seen feeding directly on vent bacteria—calanoid copepods on bacterial suspensions, vent fish on microbial mats, and the benthic polychaete *Alvinella* on *Thiothrix*-like filaments lining its tubes.

Estimates of global chemosynthetic primary production at vents is difficult. However, aerobic chemosynthesis may produce about 16×10^6 tonnes $C\,yr^{-1}$, which is about 0.1% of the photosynthetic primary production in oceanic surface waters (18.7×10^9 tonnes $C\,yr^{-1}$ and *c.* 0.02% of the total photosynthetic primary production by plants on land and sea (77.6×10^9 tonnes $C\,yr^{-1}$). In other words, although hydrothermal vents are biologically and geochemically dramatic, their contribution to global primary production appears to be minute.

CHAPTER SEVEN

THE INTERTIDAL ZONE

Tidal ranges and levels

On most coasts the tide's amplitude is large enough to expose and cover a strip of land, the intertidal zone, twice each day. The cause and variability of tides have already been discussed (Chapter 2). Here we shall consider the intertidal environment and the animals and plants living in it.

At the fortnightly spring tides the tidal range is large. The tides rise on average to the mean high water springs level (MHWS) and fall to the mean

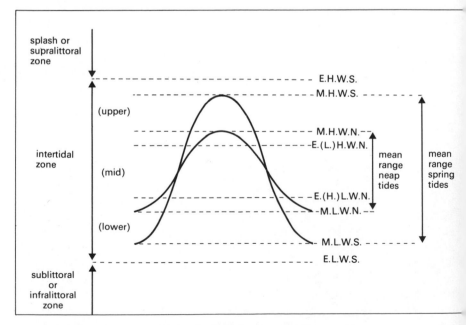

Figure 7.1 Tidal levels on the shore.

low water springs level (MLWS). During the equinoxes they occasionally rise to extreme high water springs (EHWS) and fall to extreme low water springs (ELWS) (Figure 7.1). At the intervening fortnightly neap tides, the range is much reduced. Tides rise on average to mean high water (MHWN) and fall to mean low water neaps (MLWN). During the solstices they occasionally rise only to extreme lowest high water neaps (E(L)HWN) and fall only to extreme highest low water neaps (E(H)LWN). Mean sea level (MSL) is calculated from the mean of tide heights at hourly intervals over a month or more. Ordnance Datum (OD), which is about mean sea level, is the level from which all land surveys in Britain are measured. There are two ODs, differing by a few millimetres: OD (Newlyn) and OD (Liverpool). Chart datum (CD), lying near MLWS or ELWS, is the level to which soundings on a naval chart are measured.

Supralittoral, intertidal and sublittoral zones

The intertidal zone extends from the lowest level exposed to air by tides or waves to the highest level washed by tides or waves. The supralittoral splash zone lies above the intertidal zone and is limited by the level to which sea spray flies. The sublittoral or infralittoral zone extends from the intertidal zone into deeper water. The intertidal zone is sometimes divided into the upper, mid, and lower shore or tidal zones. The limits of these zones are often difficult to define exactly. The rise and fall of the tide usually follows a sine curve; levels change slowly at high and low tide and quickly halfway between, that is, during the ebb (falling) and flow (rising) tide.

Rocky, sandy and muddy shores. Wave action. Breaker, surf, swash zones

On different parts of the coast, intertidal zones fall broadly into three types: rocky, sandy and muddy shores. Shore shape from above high to below low water is determined by the type of shore, the tidal range, and wave action. Wave action and the energy dissipated on the shore are the major factors controlling the development of shores. Waves shorten and steepen as they approach the shore (Chapter 2). They form a breaker zone in which the waves first break, a surf zone, and finally a swash zone (Figure 7.2). In the breaker zone the circular orbital motion of water particles changes to a tilted ellipse, the particle velocity in the crest increases, and the front of the crest overtakes the main wave and then collapses in front of it. In the surf zone, after the wave has broken the water is turbulent and contains many

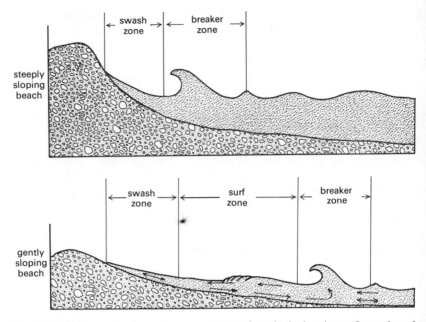

Figure 7.2 Breaker, surf, and swash zones on steep and gently sloping shores. On moderately sloping shores the surf zone is sometimes absent at high tide. Water currents in the three zones are shown in the lower diagram (modified from Ingle, 1966).

air bubbles and sand particles; the surface water moves towards the shore and the deep water away from it (Figure 7.2*B*). In the swash zone most of the waves' energy has been lost, water depth is a few centimetres, and the turbulent foamy water moves towards and away from the shore as a whole. The presence and extent of these zones depends partly on weather conditions (they will be more obvious with onshore high winds, for example), partly on the slope and nature of the shore, and partly on tidal level (Figure 7.2). Steep rocky shores have little or no surf zone, and their breaker zone is close inshore. On moderately sloping shores the breaker zone is further out, and the surf zone is usually present except at high tide. On gently sloping shores the breaker zone extends furthest from the shore, and the surf and swash zones are well developed.

Coastal currents. Longshore and rip currents. Longshore sand transport

Coastal currents may develop beyond the breaker zone and are induced by tidal and wind action (Figure 7.3). They run parallel to the beach. If waves

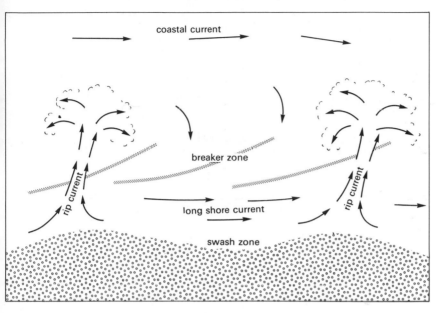

Figure 7.3 Near-shore currents (modified from Ingle, 1966). The scale depends on local conditions, but the two rip currents would probably be 10 to 100 m apart.

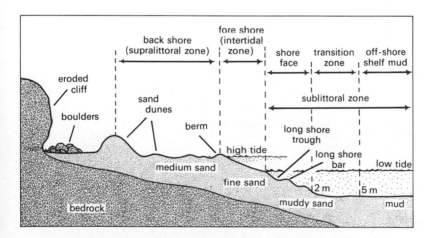

Figure 7.4 Main geomorphological zones of a sandy beach.

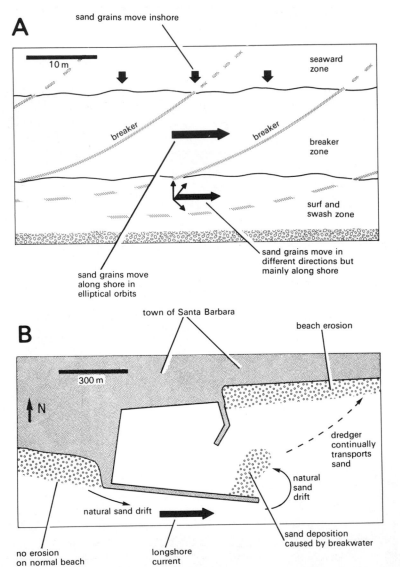

Figure 7.5 *A.* Longshore transport of sand in the near-shore currents (modified from Ingle, 1966). *B.* Interference with longshore transport of sand by the harbour breakwater at Santa Barbara, California. A dredger works continuously to remove sand from inside the breakwater and to deposit it back into the normal longshore current system at the right of the diagram. Beach erosion tends to take place at this point unless enough sand is transported by the dredger (modified from Weyl, 1970).

are not breaking exactly at right angles to the beach (Chapter 2), longshore currents form in the surf zone and follow the bars and troughs in the beach sediment (Figure 7.4). After a few tens of metres to hundreds of metres they suddenly turn seaward into rip currents. Rip currents are dangerous to swimmers. If caught in a rip current one should always swim parallel to the coast and not try to swim against the current towards the shore. Rip currents occur where a sediment trough or bar running parallel to the beach stops. Their position can change within hours, and they are thought to be caused by water piling up in the bar-and-trough system on the beach profile. Rip currents produce channels at right angles to the shoreline which extend to depths of 5 m. Longshore and rip currents pulsate, and may flow at more than 1 metre per second (about $4 \, \mathrm{km \, h}^{-1}$). Both types of currents transport sand particles: the overall effect is a mass longshore transport of sand, mainly in the breaker zone, with particles moving in a zigzag line (Figure 7.5A) (Ingle, 1966). Longshore transport of sand along the Atlantic and Pacific coasts of North America in a southerly direction amounts to between 200×10^3 and $700 \times 10^3 \, \mathrm{m}^3$ of sand per year, and so represents an appreciable movement of sedimentary material along the coast. Groynes and harbour breakwaters often interfere with this transport, which may lead to unexpected silting of harbours because the shore current pattern has been changed (Figure 7.5B).

Changing sea level. Geomorphological beach zones.

Where mean sea level is rising in relation to the land, sandy beaches develop from the weathering of rocky shores by wave action. Waves break up the rock and cause falls from cliffs. Boulders and stones are ground into gravel and sand which then form a veneer or thin layer of sediment over the original rock platform (Figure 7.4). This process is occurring on parts of the east and south coasts of Britain at present. If the sea level is falling, sand dunes and a backshore region develop and are colonised by land animals and plants. Geomorphologists divide the beach into the backshore above high tide level, the foreshore (intertidal zone), the shore face sloping down from low tide with its sediment bars and troughs, and finally the transition zone which often flattens suddenly, contains mud or muddy sand, and leads into the mud of the continental shelf (see Chapter 6) (Reineck and Singh, 1973).

There are very wide variations in these structures. For example, fine or coarse sand may occur in the backshore, mud in the offshore zone can change to sand at 10 or 20 m depth, and the slope of the intertidal and subtidal zones may vary widely.

Sandy and muddy shores. Particle size

Sandy shores are classified into coarse sand, fine sand, muddy sand and mud shores, one type grading into the next. Animals usually burrow, and plants are usually limited to micro-algae in or on the sediment (diatoms, blue-green algae). The sand grain itself constitutes a microenvironment for a very wide range of photosynthetic and heterotropic microorganisms (Meadows and Anderson, 1966, 1968; Tufail, 1985, 1987) (Figure 7.6). Broadly

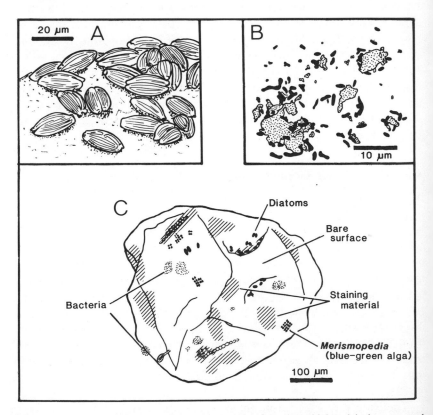

Figure 7.6 Microorganisms on sediment sand grains from intertidal and inshore coastal waters. *A*: a monospecific colony of an *Amphora* sp. (diatom). *B*: bacteria around detrital matter (stippled) (Tufail, 1985). *C*: characteristic appearance of a sand grain with a range of microbial colonies, staining material, and bare areas (Meadows and Anderson, 1966, 1968).

speaking sediments divide into gravel (20 mm–2 mm), sand (2 mm–20 μm), silt (20 μm–2 μm) and day (2μm–0.2 μm), or mixtures of the four categories (p. 113, Figure 6.5). Shape, roundness and surface texture of grains can also be measured. On most beaches the coarser grains occur towards high water. Coarse and beaches occur on open coasts where granite or similar rocks are eroded. Fine particles are washed away by waves, and the beach drains quickly as the tide falls. The beach slopes steeply, sand masses are constantly moved by wave action, and animals may be scarce. Beaches having the same grade of sediments have a steeper slope than those where there is less wave action.

Fine sand beaches are found on more sheltered coasts, and in bays and sheltered inlets where winds do not blow on shore; on these, burrowing animals are more abundant than on a coarse sand beach, and the beach slope is gentle. Sandy-mud and mud shores are completely sheltered from heavy wave action, usually have a rich burrowing fauna, and are often almost flat. They form in estuaries where the fine detritus and clay in river water precipitates on mixing with sea water, and also develop on coasts that are sheltered by offshore islands.

Origin of sand on beaches

The sand present on beaches arises from beach erosion, and from sedimentary material carried by rivers that have eroded rocks to form gravel and sand during their passage to the sea. The particles of sand on a beach therefore reflect the constituents of the beach bedrock and of the rocks over which the rivers have flowed. They are often made of quartz or felspar. Longshore transport disperses the sand from the river mouth along the coast, forming extended sandy beaches. Where a submarine canyon (Chapter 1) comes close inshore, the sand in the longshore transport system moves into the canyon under the effect of gravity. The shore line beyond the canyon is deprived of its sand supply and is often rocky. This coastal pattern is well known on the western coast of North America, where the southward moving longshore transport system is interrupted by a number of canyons. SCUBA divers have reported dramatic cascading sand rivers in some of these. It is not known how longshore transport and sand cascading affect the distribution of sand-dwelling animals.

Summer and winter beach profiles

On coarse and fine sand beaches the beach profile changes seasonally (Figure 7.7). In winter, sand is moved from the flat ridge or berm at high tide

F

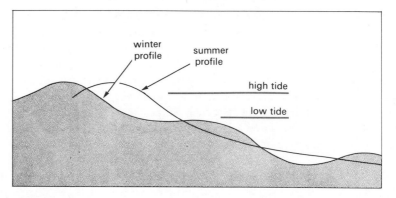

Figure 7.7 Winter and summer profiles of a sandy beach (modified from Bascon, 1960).

to bars below low tide and also to a sharper berm further inland. In summer the reverse occurs. Subtidal bars form when the ratio wave height: wave length is greater than 0.03. Once formed the bars have a marked effect on waves and inshore rip currents (see above).

Rocky shores. Exposed and sheltered sites. Animal and plant zonation

Rocky shores are usually steep and sometimes vertical. They vary from smooth glaciated granite surfaces with little water retention, through deeply fissured shales that weather easily to form rock pools, crevices and gulleys, to gravel shores strewn with large boulders.

Because they have a wide range of habitats, rocky shores in the temperate and subtropical zones are often heavily colonised by invertebrates and seaweeds. Intertidal seaweeds are less obvious in the Arctic and Antarctic, partly because freezing, ice rafting and ice abrasion limit them to the subtidal. The same is often true of the tropics, and is probably caused by excessive heat, desiccation stress, and periodic freshwater flooding during the rainy season.

On exposed coasts with heavy wave action, rocky shore communities are largely controlled by wave force. On sheltered coasts other factors, such as predation and inter- and intraspecific competition are important. Lewis (1964) has described the typical biological differences between exposed and sheltered sites in the British Isles (Figure 7.8). Animals and plants are zoned at both sites, forming broad horizontal bands on the rock. At the exposed site the black lichen *Verrucaria maura* forms the highest zone with the gastropod *Littorina neritoides*. At the sheltered site the gastropod is absent.

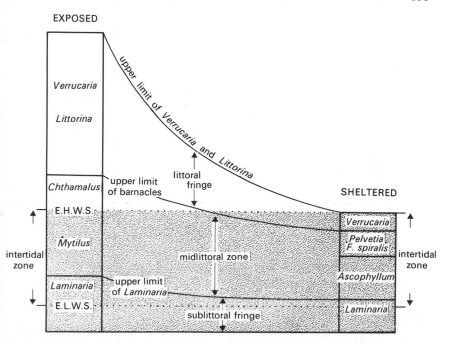

Figure 7.8 Rocky shore zonation at an exposed site (Parkmore Point, County Kerry, Eire) and at a sheltered site (a channel between the mainland and Seil Island, at the northern end of Clachan Sound, Argyll, Scotland). The intertidal zone at the exposed site is 3.4 m in extent and at the sheltered site 2.3 m. At the exposed site the *Verrucaria/Littorina* zone extends up to 10 m above spring high tide (modified from Lewis, 1964).

The splash zone at the exposed site reaches an extraordinary 10 m above MHWS, and is followed by a zone of *Chthamalus stellatus*, a highwater barnacle. Between MHWS and MLWS, the exposed site is dominated by the molluscs *Mytilus edulis, Patella vulgata* and *Thais lapillus*, with a few stunted fucoid seaweeds, while the sheltered shore is dominated by a series of seaweed bands: *Pelvetia canaliculata, Fucus spiralis, serratus,* and *Ascophyllum nodosum* (Figure 7.10). Both sites have a seaweed zone of *Laminaria digitata* at and below MLWS. The absence of mid-tide seaweeds and the enormous extension of the *Verrucaria zone* above high tide at the exposed site are almost certainly caused by wave action.

Together, these sites show the three basic rocky shore zones.

(i) Uppermost or littoral fringe: encrusting lichens (*Verrucaria*), marine gastropods (*Littorina*), isopods (*Ligia*) and a few barnacles (*Chthamalus*).

(ii) Middle or mid-littoral zone: barnacles (*Semibalanus*), gastropod limpets (*Patella*), mussels (*Mytilus*), and in northern temperate climates fucoid seaweeds (*Fucus*).

(iii) Sublittoral fringe: large brown seaweeds (*Laminaria*) or in the tropics smaller fucoids (*Cystoseira* and *Sargassum*), marine grasses and corals. There are obvious differences in the species' composition of these zones between major geographical regions, for example, between the tropics and temperate zones, and also between more local regions, for example the northern and southern coasts of Britain (Southward, 1965).

Sedentary animals such as ·barnacles and tube worms (*Pomatoceros, Spirorbis*) are fixed to the rock as adults. Many intertidal animals are mobile however; winkles (*Littorina*), limpets (*Patella*), various crustaceans (hermit crabs (*Pagurus*) and shore crabs (*Carcinus*)) move over the rocks, particularly when the fucoids float up and form a miniature forest after the tide has risen. In general, invertebrates that feed on the shore either feed on plankton by sieving water passing through or near them (barnacles and bivalve molluscs), rasp young seaweeds and other plant material from the rock surface, or nibble the fronds of older seaweeds (limpets, top shells (*Gibbula*) and *Littorina* species). The activities of the latter group largely control the growth of fucoids on sheltered rocky shores.

Animal and plant zonation on sandy and muddy shores

As on rocky shores, sandy and muddy shore animals occur in zones, although again there is much variation both locally and geographically. On sandy beaches around Britain, for example, the amphipods *Talitrus* and *Talorchestia* feed on and live in detritus, decaying animals, and drifted seaweed at high tide (Figure 7.9). The lugworm *Arenicola marina* (Annelida), the cockle *Cardium edule* (Mollusca) and many small crustaceans, such as *Eurydice, Haustorius*, and *Bathyporeia* burrow in the midtidal zone. Towards low water, burrows of the razor shell *Ensis* (Mollusca), the cockle *Cardium edule*, and in the tropics the crab *Emerita* are common. Below low tide the numbers of animals increase, and some species found on the shore may be more abundant here (*Cardium edule*). Mobile species, such as the sand eel *Ammodytes*, shrimps (*Leander*) and shore crabs (*Carcinus*) are also present.

On muddy shores the polychaetes *Nereis diversicolor, Arenicola marina*, and the amphipod *Corophium volutator*, burrow in the intertidal zone, and towards low water large numbers of burrowing bivalve molluscs (*Tellina,*

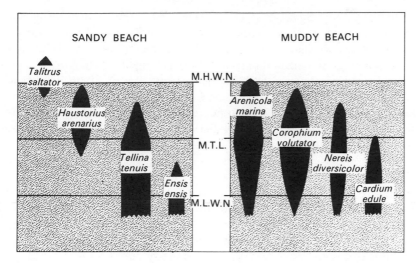

Figure 7.9 Some examples of animal zonation on sandy and muddy beaches in Britain Jagged lines at the bottom of the *Tellina, Ensis, Arenicola, Nereis* and *Cardium* distributions indicate that these species often extend well below low water.

Macoma, Cardium) occur (Figure 7.9). *Cardium* and *Tellina* inhabit both sandy and muddy shores and there are other species that occur in both habitats.

Macro-algae are not common on mud or sand. *Chorda filum, Fucus vesiculosus* and *Laminaria saccharina*, and at the top of the shore *Enteromorpha*, sometimes attach to large pebbles or boulders, while at the lower tide level the flowering plant *Zostera marina* can form extensive beds. Mobile species such as the sand eel (*Ammodytes*), blennies (*Blennius pholis*), and the shore crab (*Carcinus maenas*), are present on both sandy and muddy shores, especially when small tide pools, rocks, and clumps of seaweed are present.

Causes of zonation on shores

Studies on barnacles and seaweeds suggest that desiccation and *interspecific competition* may control the upper and lower limits of zonation respectively of many rocky shore species (Connell, 1961; Lubchenco, 1980).

The barnacles *Chthamalus stellatus* and *Semibalanus* (= *Balanus*) *balanoides* are common on rocky shores in the temporal and boreal regions of the Atlantic. *S. balanoides* has the more northerly distribution from the

Arctic to southern Spain. *C. stellatus* extends from the Shetland Islands to the Mediterranean. They overlap between Shetland and Spain. Outside their overlap *C. stellatus* grows throughout the intertidal zone, but where *S. balanoides* is present it is only found in the upper intertidal.

Field experiments and observations by Connell at Millport, Cumbrae, Scotland, proved that this was because *C. stellatus* has a greater tolerance to desiccation than *S. balanoides* and hence can live further up the shore. On the other hand in the mid tide region the more rapidly growing *S. balanoides* overgrows and undercuts *C. stellatus*, thus killing it.

Similar effects control the distribution of the red seaweed *Chondrus crispus* and the brown seaweed *Fucus vesiculosus* on rocky shores of New England, USA. *F. vesiculosus* dominates the middle intertidal and *C. crispus* the lower intertidal. Field experiments show that *F. vesiculosus* is more resistant to desiccation and so can live further up the shore, but that *C. crispus* stops it growing in the lower intertidal.

Interspecific competition on muddy and sandy shores is less obvious and may be less important. Many deposit feeding species coexist, possibly because they live at different levels below the sediment/water interface. There is also a correlation between species diversity and increasing particle size range. This makes sense because a greater range of particle sizes will allow a larger number of differently adapted species to feed. There are however some clear examples of interspecific competition. The burrowing polychaete *Armandia brevis* is common on a mud flat on San Juan Island, Washington, USA. Its abundance is not related to the physical properties of the sediment but to the space available between other species for tube construction in the sediment (Woodin, 1974).

Energy transfer, nutrient cycling, food webs

Energy transfer and nutrient cycling in the intertidal zone is complicated because there are many different types of shores and many inputs and outputs. Freshwater run-off from land may contain large amounts of nutrient from farming, and the constant ebb and flow of seawater changes the superficial and much of the interstitial water every tidal cycle. Energy loss along the beach by long-shore transport and to the sublittoral by the receding tide are also important.

On sandy and muddy beaches, microorganisms and detrital organic material are important for deposit feeders, and macro- and micro-algae are eaten by herbivorous gastropods and polychaetes. Some polychaetes practise gardening of algae in their tubes (Woodin, 1978). However, most macroalgae probably enter the food chain as detritus after death. There are marked seasonal changes on some beaches. In spring, organic detritus is deposited from phytoplankton and zooplankton. On sheltered shores in temperate climates there can also be major growths of the macro-algae *Ulva lactuca* and *Enteromorpha intestinalis*. These grow during summer, but are then rapidly broken up and destroyed by autumn storms (Hylleberg, 1977; Meadows and Tufail, unpublished observations). Masses of decaying seaweed float backwards and forwards in enclosed bays, and at Ardmore Point, Clyde Estuary, Scotland, these can be 20 or 30 cm deep at times. Polychaetes such as *Nereis brandti* and *Arenicola marina* draw fragments of algae down their tubes and into the sediment, but this only accounts for *c*. 10% of the disappearing seaweed. The rest is decomposed by microorganisms to detrital material and hence recycles nutrients into the intertidal zone.

The specific problems associated with salt marshes have received considerable attention, and are dealt with elsewhere (Chapter 8).

On rocky shores different constraints apply: herbivores largely control the level of production, species composition, and structure of the attached macro- and micro-algal community (Paine, in Goulden (ed.), 1977). For example limpets (*Patella*) and periwinkles (*Littorina*) can sometimes completely clear rocky surfaces of algae. However, grazing is usually more localised—being controlled by the spatial heterogeneity and desiccation of the rocky shore environment. Carnivores then eat the herbivores. On some shores these may be represented by one keystone species such as the starfish *Pisaster ochraceus* that dominates the wave-exposed coasts of Washington, USA (Paine, 1966). This species appears to control the whole community structure of the shore, because if it is experimentally removed the number of invertebrate species falls from 30 to one—*Mytilus californianus*. On other

Table 7.1 Ages of intertidal macro-algae (modified from Boney, 1966).

Species	Age in years	
	Average	Maximum
Fucus vesiculosus	1	3
Fucus spiralis	1.5	4
Laminaria saccharina	3	*c*. 3
Pelvetia canaliculata	2–3	3–5
Ascophyllum nodosum	12–13	13–19

shores especially in the tropics and subtropics, there may be a number of predatory carnivores none of which is dominant.

The effect of grazing by herbivorous invertebrates on the macroalgal cover of rocky shores may be related to the age and rate of growth of the macro-algae themselves (Table 7.1). For example the effect on the young stages of a slow-growing long-lived species such as *Ascophyllum nodosum* may be greater than on the young stages of a rapidly-growing short-lived species such as *Fucus vesiculosus*. The latter can probably replace its population more quickly than the former, and can therefore compensate for the grazing more efficiently.

Adaptations to intertidal conditions

Marine animals and plants on rocky, sandy, and muddy shores live in an environment which is almost one-dimensional—a thin line along the coast. They are exposed to air twice a day, they experience wide temperature and salinity fluctuations (0° to 30° C and less than 1‰ to above 35‰), and often encounter the force of heavy wave action. A few upper shore species are air-breathing; some are land forms such as the primitive insect *Petrobius*, and some were originally marine, such as the crustacean *Talitrus saltator* and the gastropod *Littorina neritoides*; but most species only breathe in sea water. All of the species living on the shore are adapted in one way or another to the changing conditions of the intertidal environment. *Arenicola*, for example, contains a haemoglobin which can take up oxygen at the very low concentrations that exist in the animal's burrow when the tide is down.

Four examples will illustrate the specialised local environments that can occur on the shore and the adaptations that animals and plants living in them have developed.

Rock pools are found on many rocky shores and are colonised by macroalgae and a number of intertidal invertebrates. Species diversity in the pools is lower in high-tide pools than in pools further down the shore. Rock pool animals fall into four categories: true rock pool species, species abundant in pools but also found elsewhere, species sometimes found in pools but usually found elsewhere, and seasonal immigrants into pools (Emson, in Moore and Seed (eds.) 1985).

Taylor and co-workers (Morris and Taylor, 1985; Taylor, 1987) have described the marked environmental changes in pools near high tide on the Isle of Cumbrae, Firth of Clyde, Scotland. Oxygen fluctuates between 400 Torr or 215% supersaturated in late afternoon and 20 Torr or 13%

saturated just before down, and carbon dioxide shows a matching but opposite trend. This is because during the day photosynthesis by algae exceeds respiration by algae and invertebrates; at night there is no photosynthesis but algal and invertebrate respiration continues unabated (1 Torr = 1/760 atm. 100% O_2 saturated sea water has an O_2 of c. 160 Torr at 10 °C).

There are also large climatic and seasonal effects. Changes in the weather can produce marked daily changes. Oxygen is significantly higher on sunny days than on cloudy days and freshwater run-off after rain in winter causes low salinity. Temperature and salinity are much higher in summer than in winter and in addition the diurnal range of temperature and oxygen is higher in summer. There is also more algal growth in summer than in winter.

In north-west Europe the glass prawn *Palaemon elegans* is one of the commonest rock-pool inhabitants, particularly towards high tide, although in non-tidal areas of Sweden it is found in sublittoral beds of *Zostera marina*. Larger specimens may migrate offshore in winter.

Palaemon elegans collected from rock pools show a number of physiological and behavioural adaptations to the rock pool environment. They maintain their oxygen consumption constant down to an environmental level of about 15 Torr—a level not often encountered under natural conditions. This is achieved by increasing the rate of water movement across the gills. Their blood also has a higher oxygen carrying capacity and oxygen affinity than that of the related species *Palaemon adspersus* which does not live in rock pools (Table 7.2), and this may be an adaptation to rock-pool life. *P. elegans* can even survive short periods of total anoxia by utilising anaerobic metabolism involving the production of lactic acid. This is interesting, because increased lactate levels in the blood apparently increase the oxygen affinity of the oxygen-carrying pigment haemocyanin. *P. elegans* also shows behavioural adaptations to low oxygen. At night,

Table 7.2 Respiratory physiology of rock pool and non-rock pool *Palaemon* species (Decapoda, Crustacea) (Weber and Hagerman, 1981; Morris *et al.*, 1985).

Species	Habitat	O_2 carrying capacity (mmol. L^{-1})	O_2 affinity (P_{50} Torr 15 °C pH 7.9)
Palaemon elegans	Intertidal rock pools	1.17	4
Palaemon adspersus	Sublittoral *Zostera* beds	0.98	16

when the oxygen level in the rock pools drops to very low levels, it moves to the edges of the pools where the oxygen content is slightly higher due to diffusion from the atmosphere into the shallow water.

The second example comes from the upper interstitial and supralittoral zone and concerns the Talitridae (Spicer, Moore and Taylor, 1987). *Orchestia gammarellus* and *Talitrus saltator* are jumping amphipods that

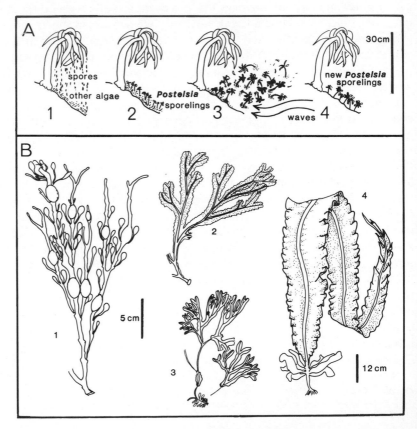

Figure 7.10 Seaweeds from North American Pacific and from European coasts. *A*: Recolonisation of surf-swept rocks by *Postelsia palmaeformis* (British Columbia, Canada). 1, Spores shed at low tide drip onto substrate and attach to other epiphytic organisms on the rocks (algae, barnacles); 2, young sporelings grow on other organisms; 3, wave action tears host organisms with sporelings from rock: 4, new *Postelsia* settlement on the bare rock (after Dayton, 1973; Carefoot, 1977; and Boney, personal communication). *B*: 1, *Ascophyllum nodosum* (Europe); 2, *Fucus serratus* (Europe); 3, *Fucus ceranoides* (Europe); 4, *Alaria marginata* (Pacific coast of North America) (after Boney, 1966; Carefoot, 1977).

live in the drift-line seaweed at the top of the shore on European coasts. Both are well adapted to their environment and can survive almost indefinitely in fully saturated air. However, they are very susceptible to desiccation, and this tends to limit them to the high-tide seaweed drift line. Both species emerge at night, and *Talitrus saltator* is known to make excursions from its burrow in the drift line towards the mid-tide level when the tide is down. It only does this at night, and is able to recognise and return to the upper shore by the slope of the beach and the outline of the top of the beach against the sky. Their osmoregulatory and respiratory abilities are well adapted to their high tide existence, but are not broad enough to allow them to colonise the land.

The third example comes from the surf-swept rock in the middle intertidal zone of the west coast of America near Vancouver (Canada). The seaweed *Postelsia palmaeformis* is a characteristic member of the community in this very high-energy environment (Dayton, 1973; Carefoot, 1977). *P. palmaeformis* is an annual species which has a unique method of spore release. Spores are shed at low tide, flow down grooves on the fronds, fall onto the substrate, and attach before the tide comes in (Figure 7.10). The microscopic gametophytes grow on algae and barnacles to produce young *Postelsia* sporelings. These sporelings are attached to the animals more strongly than the animals are to the rock, so as they grow the force of the heavy waves on the *Postelsia* fronds tears the animals from the rocks. The newly exposed bare rock is then colonised by more *Postelsia* and these sporelings remain and grow to adulthood.

Lastly, sediment burrowers are often specialised feeders. A few are carnivorous or omnivorous (*Nereis*). Many are deposit feeders and eat sediment indiscriminately as they burrow (*Arenicola*) or select certain sizes of particle (*Amphitrite johnstoni*) (Polychaeta). Some are suspension feeders passing a large volume of water through gills that filter off suspended plankton and detritus. Benthic animals living below low tide feed in similar ways. On muddy shores the subsurface sediment is often anaerobic, and burrowing animals either pass a water current through their tubes (*Corophium, Arenicola*) or extend tentacles or siphons on to or above the sediment surface (*Amphitrite, Tellina, Sabella*) (Figure 6.7).

Behavioural responses and habitat selection

Many benthic animals in the intertidal zone show characteristic patterns of behaviour and habitat selection which may explain their distribution in different sediment types, on rocks, and up and down the shore. Meadows and co-workers (e.g. Crisp and Meadows, 1962, 1963; Meadows 1964; Meadows and Campbell, 1972 *a, b*; Anderson and Meadows, 1978; Meadows and Tufail, 1986) have studied these patterns of behaviour over many years. They suggest that the distribution of animals both at a micro-level of cm and m, and at a macro-level of tens of metres to km, are largely determined by behavioural responses to the environment, not by lethal limits or extremes of physiological stress. These responses operate throughout the life cycle of mobile species, and during the larval stages for these species that are entirely sedentary as adults. Furthermore they are a general feature of all parts of the marine environment, not just the intertidal zone. Environment stress causing physiological change and in extreme cases mortality may therefore not be as widespread as is sometimes thought (Figure 7.11). Central preference curves can be viewed as being flanked by upper and lower mortality curves (lethal limits). In Figure 7.11, the *y* axis is the probability of an animal preferring (preference curve) or being killed by (mortality curve) the range of variable shown on the *x* axis. The variable might be salinity, oxygen or temperature, for example. The animals fall into three groups, *A*, *B* and *C*, depending on, say, age, sex or physiological state. Lethal limits fall outside the preference curves for each of the groups, and the onset of physiological change lies between the preference curves and mortality curves.There are arguments for and against these views, but there is undoubtedly a great deal of evidence showing that animals actively select their preferred habitat and avoid unfavourable habitats.

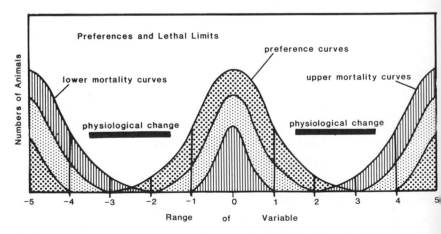

Figure 7.11 Relationship between preferences and lethal limits (mortality) (Meadows and Campbell, 1972*b*).

The following examples show how well-defined some of these behavioural responses are. Cyprid larvae of barnacles (*Semibalanus balanoides*) settle near adults on the shore. This gregarious response is dependent on the larvae recognizing a specific protein in the adult's tissues (Crisp and Meadows, 1962, 1963). The mud burrowing shrimp *Corophium volutator* is found at high densities on some muddy shores. Its behaviour shows that it is highly gregarious, responds to microbial films on sediment particles, prefers finer grained sediments, and has specific preferences for certain combinations of salinity and temperature (Table 7.3). The behavioural preferences of *Corophium volutator* help to explain the species' localised distribution in the intertidal zone.

Reise (1985) has described the interesting habitat changes of the lug worm *Arenicola marina* along a tidal gradient at Königshajen, Island of Sylt, North Sea (Figure 7.12). *Arenicola marina* has no pelagic larvae. Eggs are fertilised in the female's burrow and larvae stay there until autumn. In their first winter they move to mud patches between mussel beds at low tide where they construct horizontal tubes just below the sediment surface and feed on detritus. At this point they are 1 to 6 mm in length. In spring they

Table 7.3 Habitat preferences of the mud-burrowing amphipod *Corophium volutator*. Animals are offered a choice of two habitats as follows: fine and coarse sediment, aerobic and anaerobic sediment, sea water of two different temperatures and salinities, sediment occupied by animals and unoccupied sediment (data selected from Meadows 1964a, c; Campbell and Meadows, 1974; Meadows and Ruagh, 1981).

	Percentage of animals in:	
Habitats offered to animals	preferred habitat	non-preferred habitant
(i) Fine sediment preferences < 44 μm *compared with* 44–104 μm	90	5
(ii) Aerobic sediment preferences aerobic *compared with* anaerobic sediment *compared with* sediment	72	28
(iii) Combined temperature-salinity preferences 15 °C, 20‰ *compared with* 25 °C, 5‰	80	20

(iv) Gregarious behaviour			
		Percentage of animals choosing	
sex of introduced animals	sex of animals already in sediment	occupied sediment	unoccupied sediment
♀	♂	84	4
♂	♀	44	0

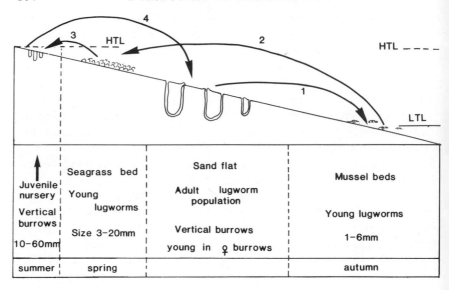

Figure 7.12 Habitat changes taking place during the life of *Arenicola marina* along a tidal gradient on a North Sea mud flat (after Reise, 1985).

move to an area of seagrass in the upper intertidal zone. Again they construct horizontal tubes, and feed on detritus and diatoms. By this time they have grown to between 3 and 20 mm. From July on, the young worms have moved to a nursery bed just below high tide. They construct vertical tubes 10 to 60 mm in length, ingest sediment, and produce characteristic faecal coils on the surface. In Autumn, the subadult worms move down the beach again to join the adult population in the mid tide region. Reise regards the low-tide and high-tide areas used by the growing worms as being safe from predators.

CHAPTER EIGHT

ESTUARIES AND SALT MARSHES

Coastline modification. Sediment load

Estuaries form where rivers flow into the sea. Their shape varies with the river flow, tidal range, the local coastline and so on. Estuaries of large rivers modify the coastline and sublittoral topography for many kilometres around by deposition and erosion of bottom sediments. For example, the Mississippi River is very broad and slow, carries a large sediment load of about 10^{11} kg per year, and has a delta of land at its mouth 10 to 20 km wide. The river sediment can be traced from the delta across 120 km of continental shelf which deepens slowly to about 50 m, and then down the continental slope to the abyssal plains.

Most estuaries have almost certainly been formed since the last glacial period about 100 000 years ago when the sea level was about 100 m lower than today (Levinton, 1982), and on the Atlantic coast of North America, most are younger than 5000 years. Estuaries are therefore transitory ecosystems on a geological time scale and may be formed and reformed following successive rises and falls in sea level. Estuaries can form from drowned glacial valleys or fjords, drowned river valleys, and basins formed by tectonic activity (movement of crustal plates). They can also form in bays that develop behind beach bars.

The history of the Mississippi delta complex shows how rapidly some estuarine environments can change (Figure 8.1). The origins of the Mississippi delta can be traced to the area around Cairo, Illinois, in the Cretaceous era ($> 65 \times 10^6$ yr b.p.). Since that time it has advanced about 1600 km to its present position. Over the past 6000 years, the delta has shifted eastwards and then westwards of the present delta several times (Kolb and Van Lopik, in Shirley (ed.), 1966; Frazier, 1967; Press and Siever, 1982). The oldest complex was below the present area of Atchafalaya Bay and the mouth of the Atchafalaya River (c. 6000 to 4000 yrs b.p.). The complex then swung eastwards to lie under the New Orleans area (c. 4000

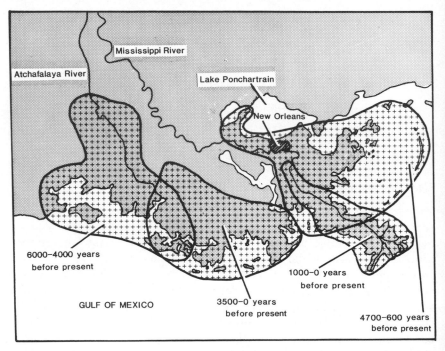

Figure 8.1 Changes in the Mississippi delta complex over the past 6000 years (after Kolb and Van Lopik, in Shirley (ed.) 1966; Frazier, 1967; Press and Siever, 1982).

to 600 yrs b.p.). Very recently (1945–1955), there were indications that the delta was beginning to move westwards again from its present site towards the Atchafalaya River, and major construction works were undertaken by the US Army Corps of Engineers.

Patterns of estuarine circulation, bores

River water progressively mixes with sea water as it approaches the outer estuary, while sea water moves in and out of the estuary with each tidal cycle (Figure 8.2). A volume of water equal to the tidal prism (the difference in volume between high and low tide in the estuary) is exchanged with the sea every tidal cycle. This exchange is superimposed on the outflow of river water. The tidal flow moves out fastest at mid-ebb tide and in fastest at mid-flood tide. The total mid-ebb flow is greater than the total mid-flood flow because of the outward flowing river water. At high and low tide there is no tidal current and the only current is that of the river.

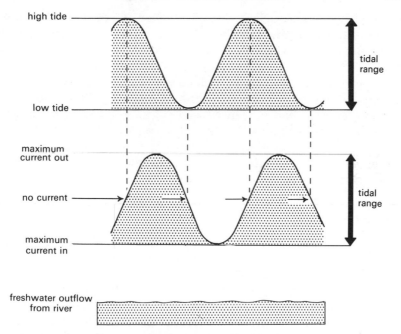

Figure 8.2 Tides in estuaries. *A*. Tidal range in relation to *B*, tidal current in and out of an estuary. *C*. The relatively constant freshwater outflow from the river.

In some rivers local topographical and tidal conditions produce larger tides in the estuary than elsewhere. Estuaries with a steadily rising bottom and a funnel shape have a rising tide that is marked by an advancing wave of water, the *bore*, moving rapidly upstream. The bore on the River Severn in Britain is 3 m high, while that on the Chiang Tang Kian in Eastern China is 8 m high. Bores travel at a few kilometres per hour.

Water circulation within estuaries depends on tidal range, vertical mixing between fresh and sea water, and the bottom topography. In general, river water flows out on top of the denser sea water which is moving in and out with the tide, while there is a layer of brackish water between the two (Figure 8.3). Currents are fastest in mid-stream, while friction reduces their speed near the sides and bottom of the channel.

Water flow and mixing in estuaries have been divided into four types by Cameron and Pritchard (1963) (Figure 8.3). In type A, the tidal range (inflow and outflow of sea water) is small, and the outward flow of fresh water from the river is large. A wedge of sea water lies below the outward-flowing river water, and there is little vertical mixing between the two. In type B, the tidal range is larger, and the inflow and outflow of sea water exceeds the outward flow of fresh water. There is more mixing between the fresh water and sea water, and the Coriolis force causes the fresh water to flow outwards on the right-hand side of the estuary in the Northern Hemisphere and on the left in the Southern Hemisphere (looking downstream). As a result the freshwater saltwater boundary is tilted. In type C, the freshwater overflow is reduced, and is greatly exceeded by the tidal inflow and outflow. Fresh water again flows on the right hand side of the estuary, but here the freshwater saltwater boundary is almost vertical. Type C estuaries are often wide. In type D, the tidal flow is large, fresh water and sea water mix completely, and the estuary itself is normally narrow.

Sediment formation

Many estuaries have a sand or mud bar at their entrances which reduces water flow. Bars are formed by longshore sediment drift, and by precipitation of suspended river detritus as it meets sea water. Within the estuary, detritus is deposited by a similar process on the bottom and banks. In general, sudden extreme conditions are more important in sediment formation than the average; one heavy river flood discharges more sediment in days than is sometimes produced in a year. Recent studies of the Thames Estuary in Britain have shown that the development of mud banks and transport of materials is very complex. Topographical features, current flow, and the building of new berths and quays all affect the process, and the river may swing to and fro across the estuary from year to year. The processes of sediment deposition and erosion appear to depend in a complex manner on various physical parameters of the sediment (shear strength, moisture content, permeability, viscosity, thixotropy, flocculation). Flocculation, which is a reversible phenomenon, begins at about 4‰ salinity, is increased by high temperatures and also by carbohydrates in solution, and decreased by the presence of proteins. In slack water, flocculation forms fluid mud layers above the bottom which slowly consolidate. As the tidal current increases some or all of the mud may be eroded. Estuarine sedimentation averages about 0.2 cm per year.

Seasonal effects on water flow and sedimentation

There are considerable seasonal and topographical differences in river flow in estuaries and these are obvious to anyone who lives near an estuary. In temperate climates heavy rain and snow in winter cause a massive increase in freshwater flow and sediment load from land erosion. This moves isohalines (lines joining points of equal salinity) seawards. The effect is marked in small estuaries such as the Newport River estuary in North California, USA, but minimal in a large estuarine system such as Chesapeake Bay, USA (Levinton, 1982).

Sedimentation and sediment transport also change seasonally. The resuspension of sediment by the incoming tide is sometimes more marked in summer than in winter and in some estuaries leads to a layer of turbid water metres thick immediately above the bottom. This summer turbidity maximum is partly caused by bioturbation and faecal pelletisation which are more abundant in summer. Biological activity produces a rougher sediment-water interface which then erodes at a lower water velocity (Rhoads, Yingst and Ullman, in Wiley (ed.), 1978).

Nutrient input, output and cycling

Estuaries are very productive for two main reasons. Dissolved nutrients enter from the river and land, and nutrients recycle between the overlying water and the biologically active sediments.

Input from rivers can be very large, either as dissolved nutrients—from agricultural land for example—or as particulate material. Much of the dissolved material precipitates as the salinity increases, and this produces the rich intertidal mud flats of many estuaries. This input can be continuous or episodic, for example after storms or heavy rain, and can lead to transient phytoplankton blooms. Nutrients also enter estuaries from the sea in bottom water. This is the water that flows into an estuary along the bottom to compensate for the surface outflow of low salinity water (Figure 8.3).

Nutrients recycle between the water and sediment as follows. Intertidal and subtidal sediments in estuaries are usually very muddy, and contain high numbers of microorganisms and burrowing invertebrates. Burrows increase the sediment/water interface, and microorganisms at the interface are active in mineralising organic material that settles to the sediment. Microbial mineralisation breaks down complex organic components to carbon dioxide and water, and releases many nutrients including nitrate and phosphate. This nutrient regeneration by microorganisms is an important part of nutrient recycling. The ventilation of burrows by infaunal invertebrates will help to disperse the regenerated nutrients into the overlying water where the horizontal and vertical mixing of the water by estuarine flow will then mix the nutrients into the estuarine ecosystem.

Salt marshes (p. 175) play a large part in productivity and nutrient flow in

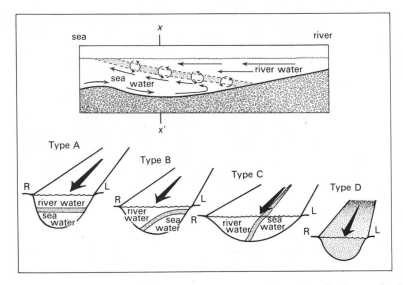

Figure 8.3 *A*. Longitudinal section down an estuary showing the wedge of fresh water flowing out over the sea water, and the mixed layer of brackish water (dotted) between the two. *B*. Cameron and Pritchard's (1963) four types of estuary. Cross section at *x–x'* in the upper diagram, looking upstream. The large black arrows show the direction of river flow. River flow tends to be reduced as one moves from type A to B to C to D.

some estuaries as they produce large quantities of detritus. Algal production and phytoplankton grazing can also have major effects. For example, in large open basin estuaries such as St Margaret's Bay, Nova Scotia, *Laminaria* forests play a major role in estuarine cycling.

Some estuaries may export large amounts of dissolved nutrients and particulate organic matter into nearshore waters, and this may increase continental shelf productivity nearby. However the processes are complex and will differ from place to place. For example, near the coast of Georgia, USA, high plankton productivity is related to several factors. The continental shelf waters in this area receive nutrient-rich estuarine waters from rivers, and nutrient-rich deep waters from the edge of the continental shelf. Nutrients are also regenerated from the underlying sediments. On the other hand in some areas estuarine input may only account for 5% of the total dissolved N in nearshore shelf production (Haines, in Cronin (ed.) 1975).

Freshwater, brackish-water and marine animals and plants and their overlap

As the salinity increases down an estuary, the number of freshwater species of plankton and benthic animals and plants decreases, while the number of seawater species increases (Figure 8.4), but there are many variations depending on local conditions.

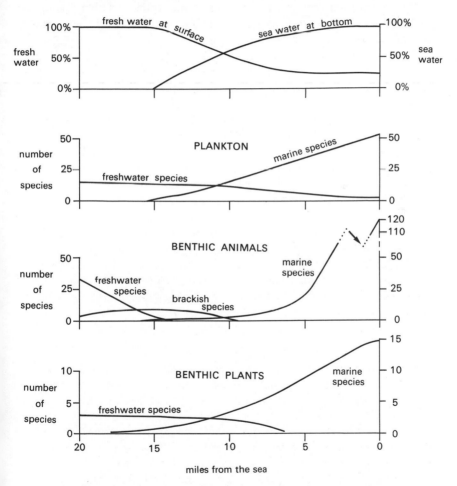

Figure 8.4 Salinity, plankton, and benthic animals and plants in the Tees Estuary, Britain (modified from Alexander, Southgate and Bassingdale, 1935).

Intertidally and subtidally the estuarine environment is a more difficult habitat than an open coast. Rapidly changing salinity over the tidal cycle produces considerable physiological stress. These effects are larger in smaller estuaries. Fresh water changes to sea water in the small Pocasset River, Massachusetts, east coast, USA, over a tidal cycle. In Chesapeake Bay, Virginia, east coast, USA, however, which is a large estuarine system, the changes are more gradual, and more affected by seasonal than by tidal cycles. The former type of estuary is sometimes called a daily estuary, and the latter a seasonal one, and these differences will affect the type and abundance of organisms that live in them.

The salinity range 5 to 8‰ is termed a *critical salinity range*, because in it Ca/Na and K/Na ratios increase rapidly with decreasing salinity. The critical salinity range is important in determining the upstream limit of marine and estuarine bivalve gastropods and the downstream limit of freshwater bivalves and this is probably true of other groups as well. Species diversity is usually lowest in the critical salinity range, and the few species remaining are often amazingly abundant. In Chesapeake Bay, oysters (*Crassostrea virginica*) and blue crabs (*Callinectes virginica*) are sometimes so abundant that they almost carpet the sediment, and in the Clyde Estuary, Scotland, some mud shores contain very large numbers of just three burrowing species: the shrimp (*Corophium volutator*) ($\leqslant 10^4 \, m^{-2}$), the ragworm (*Nereis diversicolor*) ($\leqslant 3 \times 10^3 \, m^{-2}$) and a tubificid oligochaete (*Tubifex costatus*) ($\leqslant 5 \times 10^3 \, m^{-2}$). Other marine and sand-dwelling species can also penetrate up estuaries, for example *Cerastoderma* (= *Cardium*) *edule, Arenicola marina Macoma baltica, Hydrobia ulvae* and *Mytilus edulis* in Britain (Green, 1968). These more marine species are, however, limited to higher salinities further down the estuary.

Broadly speaking, estuarine benthic animals can be classified as follows.

(i) Stenohaline marine species live in the sea and also penetrate estuaries to a salinity of 30‰ (*Tellina tenuis, Cardium edule* in Europe).

(ii) Euryhaline marine species also live in the sea, but penetrate further up estuaries to a point at which the salinity is well below 30‰ (*Carcinus maenas, Littorina littorea, Corophium volutator* in Europe).

(iii) Brackish water species, which are often related to marine forms, never occur in the sea and penetrate well below 30‰ (*Sphaeroma rugicauda* (Isopoda), a marine relative of *Sphaeroma serratum*, in Europe).

(iv) Certain freshwater species have penetrated into brackish water (*Asellus aquaticus* (Isopoda), oligochaetes and culicid larvae (Diptera)).

(v) Certain terrestrial species penetrate from the land (*Bembidion laterale* (Coleoptera) in Europe, which eats *Corophium volutator*).

Analogous schemes can be produced for freshwater and marine plankton, except that the freshwater plankton is killed as it enters the high salinities; there are also series of species down estuaries which are limited to particular salinity ranges.

There are interesting changes in the gene frequencies of some marine species at the mouths of estuaries and in other brackish-water areas, such as the entrance to the Baltic Sea (Levinton, 1982). This may lead to rapid genetic divergence in these areas. Populations of the eel pout *Zoarces viviparus* and the mussel *Mytilus edulis* show large changes in polymorphic genetic loci coding for enzymes over short geographical distances (Levinton and Lassen, in Battaglia and Beardmore (ed.), 1978).

Birds

Other animals to use estuarine environments are migrating salmon *Salmo salar*, eel *Anguilla anguilla*, and the many bird species that feed on

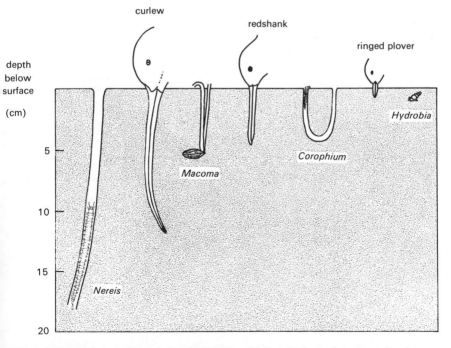

Figure 8.5 Estuarine waders and their food. Birds with long beaks, such as the curlew, have access to much deeper-burrowing species than birds with short beaks such as ringed plover (modified from Green, 1968).

intertidal organisms. Some birds such as gulls are resident, while others such as the waders overwinter on estuaries in northern temperate climates and move inshore or migrate in summer. They may include scavengers (gulls), fish eaters (cormorants, herons), omnivores (ducks, swans), mud and sand fauna predators (waders) and specialist feeders, such as flamingoes in the tropics. The bill length of waders often limits the species of burrowing invertebrates that they can catch (Figure 8.5)

Food chains and food webs

Food chains in estuaries, more accurately called *food webs* (Figure 8.6), are complex and not fully understood. Benthic deposits and suspension feeders take in food from the following sources:

  (i) Bacteria and phytoplankton
 (ii) Organic detritus from *Zostera* near estuarine mouths
(iii) Organic detritus from *Spartina* patches in salt marshes
 (iv) Organic detritus from benthic algae
  (v) Plankton-deposited river detritus.

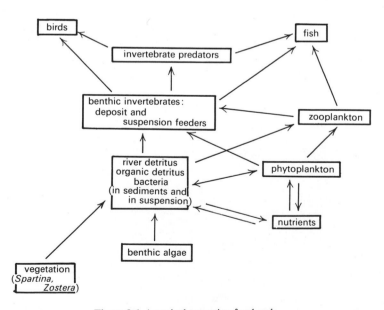

Figure 8.6 A typical estuarine food web.

The deposit and suspension feeders are eaten by invertebrate predators (*Nereis, Carcinus*) and these in their turn by fish (*Gobius microps*) and by birds, while both birds and fish also eat zooplankton (Figure 8.6).

Estuarine nurseries and feeding grounds for fish

Many species of commercial fish spend their juvenile stages inshore and in estuaries (McHugh, in Lauff (ed.), 1967). In Europe the young of herring *Clupea harengus*, and many flatfish species, are often caught well up estuaries, and on the east coast of North America the juveniles of most commercial fish use estuaries as feeding grounds. For instance the menhaden *Brevoortia tyrannus* is an important pelagic fish along the East coast of North America whose adults spawn offshore and whose juveniles feed in the Chesapeake Bay estuarine system. The juveniles are adapted to the flushing effect of the ebb tide. On the outgoing ebb tide, young fish swim near the bottom in the boundary layer where currents are less; on the incoming flood tide they move up into the water column.

Salt marshes

Salt marshes develop where mud or sandy mud flats are colonised by rooting plants, and usually occur in the upper intertidal zone of estuarine areas (Figure 8.7). They develop in detrital and silty sediments behind sand or shingle spits that have been deposited by longshore transport. They are also common in estuaries where silt is regularly deposited and wave action

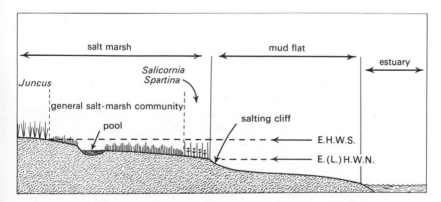

Figure 8.7 Transect across a temperate-climate salt marsh (Britain), from the land on the left of the diagram to the estuary on the right, at about mid-tide level (modified from Green, 1968).

is low. Where sea level is rising, sediment several metres deep is sometimes found. A mature salt marsh has creeks and drainage channels reminiscent of a small mangrove swamp; these channels allow tidal water to move through the marsh.

Salt marshes are common on the Atlantic coast of North America, where there are about 6000 km^2, and on some coasts in Europe. In North America the area between MTL and MHW is dominated by *Spartina alterniflora*. Above this zone are *Juncus, Distichlis spicata, Spartina patens* and *Salicornia*. The flora is more variable in Europe with significant differences between North Sea and Atlantic marshes. The Atlantic marshes are dominated by the grasses *Puccinella* and *Festuca* and are sometimes grazed by cattle and sheep. In the North Sea, the primary colonisers are the green alga *Enteromorpha*, the cord grass *Spartina* (about fifteen species in temperate waters) and sometimes the flowering plant *Zostera nana*. Plant zonation is often marked (Figure 8.7). In Britain the sea rush *Juncus martimus* may be submerged by the tide for only a few hours a month, while the cord grass *Spartina* and the glasswort *Salicornia stricta* are submerged every day. On sandier marshes the grass *Puccinellia maritima* may be dominant. A general saltmarsh plant community often develops, which includes thrift *Armeria maritima*, sea spurry *Spergularia salina* and sea lavender *Limonium*, amongst others. Salt marshes are good feeding grounds for estuarine birds since they are colonised by many estuarine and terrestrial invertebrates. At their seaward side they are often bounded by a salting cliff *c*. 1 m high.

Salt marshes may sometimes have a mangrove belt along their seaward fringe. For instance in Westernpoint Bay, near Melbourne, Australia, the salt marshes are separated from the sea by a fringe of white mangroves.

Salt marsh production, N$_2$ fixation, recycling, export

Salt marsh production can be dominant factor in some estuaries. For example, a salt marsh in Georgia, east coast, USA, is the main source of production for estuarine animals because the adjacent estuary is so turbid that there is little primary production by phytoplankton.

In salt marshes on the east coast of North America net primary production above ground varies from *c*. 250 g C m^{-2} yr^{-1} in the Gulf of St Lawrence (46° N) to *c*. 500 g C m^{-2} yr^{-1} in Texas (27° N). The range is broadly similar in Europe. Salt marshes in areas of greater tidal range are usually more productive, and productivity is usually higher on those parts of a marsh that are covered daily by the tide.

Salt marshes can export large amounts of dissolved nitrogen into the estuarine ecosystem (see below), but plant growth is often nitrogen-limited (Mann, 1982). N_2 fixation provides the answer to this apparent contradiction, and takes place by two routes (Whitney et al., 1975; Patriquin and Mclung, 1978). Blue-green algae on the sediment surface fix between 0.5 and $20 \, g \, N \, m^{-2} \, yr^{-1}$, and nitrogen-fixing bacteria on and within the underground stems and roots of Spartina fix about $10 \, g \, N \, m^{-2} \, yr^{-1}$ in the plant's rhizosphere. Nitrogen fixation is inhibited by combined nitrogen in solution such as urea and ammonia, and because of this effect sewage dumping near salt marshes reduces marsh productivity. Input of nitrogen into salt marshes occurs mainly in ground water run-off from the land. This provides twenty times more than is brought in by rain. This and the two routes of nitrogen fixation may mean that nitrogen-limited plant growth is not as important as once thought.

Little of the plant production is consumed by herbivores while the marsh plants are alive. Most consumption occurs after death. Dead Spartina tissue is rapidly and efficiently decomposed by fungi and bacteria to soluble organic material and then incorporated into microbial biomass. In some

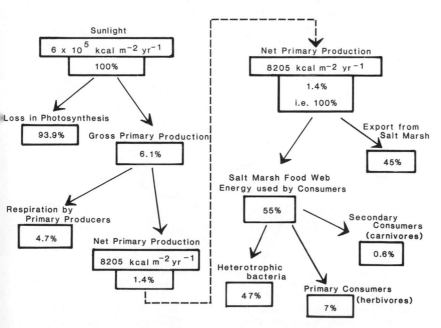

Figure 8.8 A Spartina salt marsh energy budget, Georgia, USA (after Teal, 1962).

areas the soluble organic material is taken up within the marsh as nutrients by green plants, and the microbial biomass and detritus is eaten by marsh invertebrates; thus organic material from the *Spartina* re-enters the detrital food web within the marsh. In other areas a large proportion (> 30%) is exported to the estuary or sea.

The relative amounts of organic material recycled within and exported from salt marshes depends on locality and marsh topography (Mann, 1982). The energy budget of a *Spartina* marsh in Georgia shows the details of energy flow clearly (Figure 8.8) (Teal, 1962). 6×10^5 Kcal m^{-2} yr^{-1} radiant energy is available to salt marsh plants, but net primary production only fixes 1.4% of this (8205 kcal m^{-2} yr^{-1}). However, a massive 45% of the net primary production is then exported to other parts of the intertidal zone and to the subtidal. The remainder is respired by bacteria and primary and secondary consumers.

Odum and Smalley (1959) have conducted a comparative study of the relative importance of detritus and herbivore food webs in salt marshes. They measured energy flow (kcal m^{-2} d^{-1}) through the herbivorous

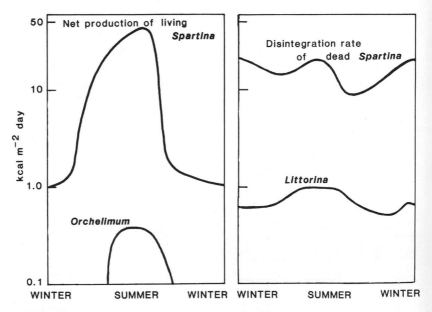

Figure 8.9 Energy flow in the herbivorous grasshopper *Orchelimum fidicinium* which feeds on living *Spartina*, and in the deposit-feeding gastropod *Littorina irrorata* which feeds on detritus formed from dead *Spartina* (after Odum and Smalley, 1959; Levinton, 1982).

grasshopper *Orchelimum fidicinium* which eats living *Spartina*, through the gastropod *Littorina irrorata* which eats detritus, and through *Spartina* itself (Figure 8.9). Energy flow through the grasshopper peaked during summer when net production of living *Spartina* was maximal, while energy flow through the gastropod was fairly uniform throughout the year in accord with the continuous disintegration of dead *Spartina*—its food. This may mean that in general, regular inputs of organic material into detrital food webs have a buffering effect on populations of animals within detrital food webs on the shore.

CHAPTER NINE

FISHERIES AND FARMING

Scale of the world's fisheries

At the present time, between 60 and 70 million metric tons of sea foods are harvested annually of which 90% are fish (1 metric ton = 1 tonne = 1000 kg) (Figure 9.1). Sea food is perhaps the only world food whose yield is increasing more quickly than human population growth. The world's fisheries are based on only a few of the 20 000 species of marine fish, and three major groupings of developed countries dominate the world market. The western European nations have the most diversified and oldest fisheries. These fisheries are linked closely to the North American and Australasian industries by their technology and also by their approach—which is basically capitalistic. In contrast, the socialist state-run industries of Russia and the Eastern European countries are centrally planned and concentrate on distant water and remote base operations. Thirdly, Japan, the world's greatest fishing nation, has recently expanded its interests on a global scale (Figure 9.2).

Markets are conservative, being largely determined by consumer choice, and this is often the limiting factor in new food developments. In addition, fish processing relies mainly on a few well-tried processes such as freezing, salting, smoking and canning, as well as on fresh fish. There have, however, been marked changes in preparation and consumption over the years (Holt, 1969; Couper, 1983). Since 1948, pelagic clupeoid fish (herrings, pilchards, anchoveta) have risen 33% to 45%, while bottom-living gadoids (cod, haddock, hake) have fallen 25% to 15% although increasing in absolute terms. Since 1938 frozen fish have increased dramatically from 0% to 20%, high-value canned products such as tuna and salmon have become important, and one-third of the world catch—mainly shoaling pelagic species—is now processed to meal or oil for fertilisers or animal food.

The prognosis for the developed countries is reasonable, with some predicted increase to the year 2000. The developing countries are likely to

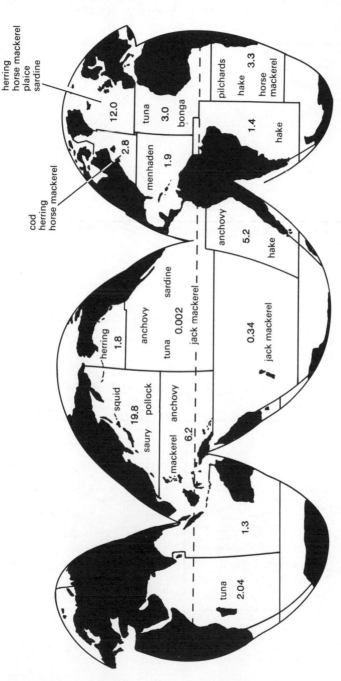

Figure 9.1 The main marine fisheries of the world. The figures are catches in millions of tonnes (1 tonne = 1000 kilograms) for all the commercially caught fish in each of the FAO fisheries areas for the 1970s. They are to be regarded as approximate only. The names are those of the more important commercial species (modified from Anon.,1972; Herring and Clarke, 1971; Couper, 1983).

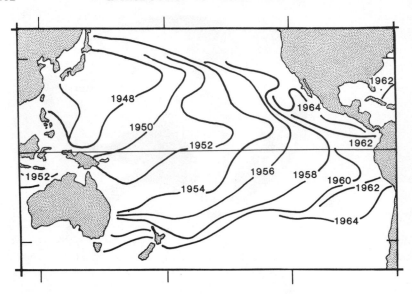

Figure 9.2 Map of the expansion of the Japanese long-line tuna fisheries since 1948 (after Kamenaga, 1967; Herring and Clarke, 1972).

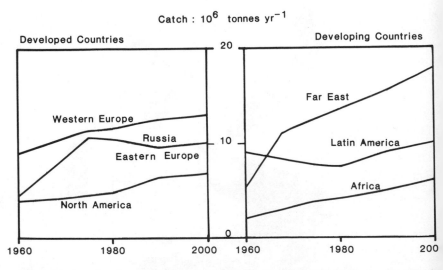

Figure 9.3 Actual and predicted world catches of finfish 1960–2000 in the developed and developing countries (after Couper, 1983).

fare slightly better because at the present time their resources are less exploited (Figure 9.3).

The fishing industry itself is a complex system involving the marine environment, people and equipment. It can be considered as a resource (the population of fishes) which is acted upon by fishing fleets and their associated catching technology. Once in port, the ships use the port infrastructure and organisation. Finally the fish are processed and distributed. This whole process interacts with the marine and coastal environment, and may include political, legal economic aspects and social pressures. These in turn can be influenced by management at a local, national or intergovernmental level.

The main countries that rely on their fishing industries to produce seafood are Russia and Poland, Scandinavia, Spain, Portugal and Japan. Traditional fisheries are also important in South-East Asia and West Africa. The main exporting countries are North America, Western Europe, and Japan.

Most of the world's fishing vessels are small boats used in developing countries. The number of vessels in commercial fleets of developed countries is comparatively small. These are usually classified by gear (seiners, trawlers) species (pole and line tuna vessel) or operating range (near/off shore, distant water).

Near-shore fisheries are usually in a country's coastal zone, and include shellfish and salmon. *Continental shelf fisheries* (inshore fisheries) operate for demersal and pelagic fish to about 500 km ($\simeq$ 300 miles). They work mainly within the exclusive economic zone. *Middle* and *distant water fisheries* include the north-east Atlantic trawling industry for cod. Finally there are the *distant water fisheries* that operate from remote floating bases or factory ships (Russia, Poland, East Germany, Japan).

Fisheries regulation is practised by most of the developed countries, and the following methods can be used.

   (i) Minimum mesh size of nets (lets small and younger fish free)
  (ii) Minimum legal sizes of fish that can be landed
 (iii) Decrease in the number of fishing vessels
  (iv) Control of the building of fishing vessels
   (v) Control of the amount of time that fishing vessels spend fishing
  (vi) Fishing only allowed within a restricted fishing season
 (vii) Fishing not allowed in certain areas (e.g. spawning areas).

However, international disagreement or non-cooperation often cause difficulties. Added complications arise from arguments between nations about the extent of territorial fishing waters and exclusive economic zones (EEZs) (p. 219) around islands and continents.

Geographical areas of fisheries

Geographically, the north-east Atlantic and north-west Pacific produce the highest catches of fish (Figure 9.1). The North Atlantic contains the oldest fisheries, and many have been overfished for some while (Table 9.1). The main fishing fleets in the north-east Atlantic are European, Scandinavian, Icelandic, and Russian, and in the north-west Atlantic are Canadian, United States and Russian. The most important fish are haddock, herring, cod, plaice, sardine and mackerel (Figure 9.4). The sand eel and red fish could be more heavily fished in these areas, while the blue whiting and argentine are abundant west of the British Isles. Towards the equator, demersal fish such as cod and plaice disappear, and pelagic sardine, horse mackerel and menhaden are fished. In the tropical Atlantic many pelagic species occur—hake, bonga, tunny, and bream. The main fishing fleets here are the United States, Venezuelan and Mexican fleets in the west central Atlantic, and Russian, Senegalese, Nigerian and Spanish fleets in the eastern central Atlantic. In the southern Atlantic there are fisheries for hake, pilchards and horse mackerel, and the main fishing fleets are Brazilian and Argentinian in the south-west and Russian, South African and Namibian in the south-east.

In the Pacific Ocean, the most important areas are the continental shelf and surrounding seas off Japan (in other words the north-west Pacific) where the major fishing fleets are Japanese, Russian and Korean, and the most important fish are anchovy, mackerel, saury and squid. The north-

Table 9.1 Exploitation of world fisheries. North Atlantic fisheries. Selected data (after Holt, 1969).

Area	Fish	Dates since which stock has been overfished
North Sea	Plaice, haddock	1890–1910
South and East Iceland, Faeroe Islands	Plaice, haddock, cod	1930s
North Sea	Herring	1950s
North-West America (Labrador, Newfoundland, Greenland)	Cod, ocean perch	1956–1966

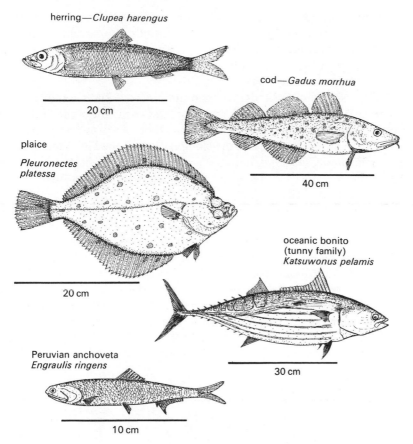

herring—*Clupea harengus*

20 cm

cod—*Gadus morrhua*

40 cm

plaice

Pleuronectes platessa

20 cm

oceanic bonito
(tunny family)
Katsuwonus pelamis

30 cm

Peruvian anchoveta
Engraulis ringens

10 cm

Figure 9.4 Some examples of commercially caught fish.

east Pacific is fished by the Canadian, United States, Russian, and Japanese fleets but the fishing effort is less than in the north-west Pacific. The waters off Chile and Peru—the south-east Pacific—are interesting because the fisheries are based mainly on one species, the Peruvian anchoveta, *Engraulis ringens* (Figure 9.4). The main nations fishing this area are Peru and Chile. The fishery is a major producer of fish meal and oil. The anchoveta itself eats phytoplankton, so this is a very short and highly efficient food chain compared to the more usual longer ones (Figure 4.4, p. 75). The anchoveta fishery depends on the constancy of a southerly wind causing an offshore surface current (Figure 2.13, p. 28) which produces upwelling of nutrient-rich water and hence massive phytoplankton growth. Every five or ten years the wind fails and so the offshore current ceases. This inhibits the upwelling, phytoplankton growth is greatly reduced, and the anchoveta population falls catastrophically. The anchoveta fishery has contracted dramatically since about 1970.

The fish stocks of the Indian Ocean are less well known and not fully exploited. Most of the fishing is by small boats near the shore. The main fishing nations in the West Indian Ocean are India, Pakistan, Sri Lanka and some of the Arab states. The main fishing nations in the East Indian Ocean are Burma, India, Thailand, Indonesia and Bangladesh. There are probably exploitable fish stocks in the upwelling waters in the Gulf of Aden and off Somalia and Southern Arabia.

Oceanic fisheries, which are found in all the oceans, are based on migratory tuna and salmon. There are six major species of tuna found between 35° S and 70° N. The most important are the skipjack, yellow fin, albacore and bigeye. In the Pacific the skipjack fishery is in the central west region, off the north island of New Zealand, and also on the coast of Central and North America. The species spawns in the central west area and undertakes a series of complicated migratory loops north, east, and south.

Pacific salmon and North Atlantic salmon migrate large distances mainly in an east/west direction (Figure 9.5). Both spawn in fresh water. The Pacific salmon contains five major commercial species—pink, chum,

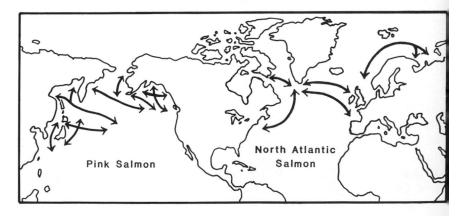

Figure 9.5 Migrations of Pink Salmon in the Pacific and North Atlantic Salmon in the Atlantic from their spawning grounds in fresh water to their deep-water feeding grounds (modified from Couper, 1983).

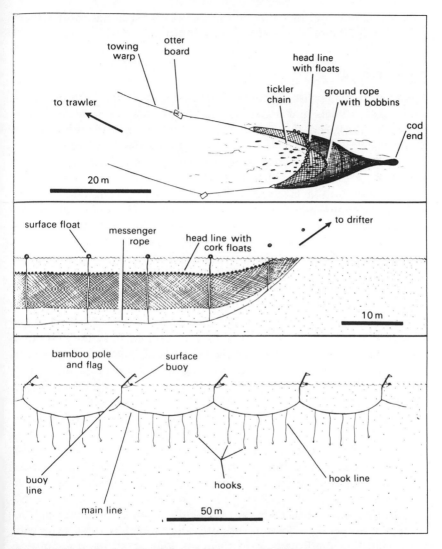

Figure 9.6 Methods of catching fish used by commercial fishing vessels.
Top: The commercial bottom trawl, commonly used on continental shelves for demersal fish (haddock, cod).
Middle: A drift net used over continental shelves for catching pelagic fish (herring).
Bottom: Long line used by the Japanese for catching tunny over the deep oceans (modified from Herring and Clarke, 1971).

sockeye, chinook and coho. All salmon are heavily fished, and most of them suffer from pollution in coastal regions and rivers.

Fishing methods

Methods of catching fish depend on fish behaviour and there are, of course, many kinds of nets and hooks. On the continental shelf, demersal fish (plaice, haddock, cod) living on or near the bottom are caught by bottom trawl (Figure 9.6). Trawls are hauled at 1 to 3 knots from side or stern trawlers. Recently, commercial pelagic trawls have been developed that fish in mid-water. Long lines with individual hooks at a few metre intervals are used when the bottom is too rough or too deep for trawling, and are often used on the continental slope to catch large cod and halibut. Pelagic fish over the continental shelves are caught by drift net (e.g. herring). Drift nets are left to lie for four to six hours overnight, and catch fish that migrate to the surface. The fish cannot see the vertical wall of netting and so are caught by their gills. Some pelagic fish are caught at night by lights and by suction devices or electric stunning although these methods are less commonly used. Recently, factory ship methods have been used on the Scottish herring grounds by continental countries and are depleting the local herring stocks to an uneconomic level. The Russians, for example, have used 60 purse seiner and 3 mother ships off the Shetland Islands and the west coast of Scotland to fish for herring. (In purse seining, a net is paid out from a fishing vessel—the purse seiner—to form a circle of vertical netting at the sea surface, which is then hauled in from both ends of the circle.)

Pelagic fish over the deep oceans are caught by a number of methods. Pole and line fishing (rod and line), practised successfully by the Americans and Japanese in the Pacific, catches young skipjack and tunny (Figure 9.4). Live bait (anchovy or sardine) is thrown overboard to attract fish and these are then caught by a rod and line operated by one or two men from the side of the vessel. The fish although young are 1 to 2 m long. Purse seining is also used. Here a wall of netting is shot to encircle a school of tunny, and the bottom of the net then closed by a bottom line. Dolphins often swim with tunny and lead the fishermen to them. As long as all the dolphins are encircled the tunny go with them. Every effort is made to release the dolphins after the tunny have been caught. The most important method of tunny fishing, however, is the long line used by the Japanese who catch over 50% of the world's tunny (Figure 9.6). Lines are fished in loops (or baskets) which are about 300 m long. A Japanese tunny long-line vessel may fish and haul 350 to 400 loops or baskets a day, equivalent to an astonishing 120 km of line.

Fish migration, North Sea herring stocks

We can now turn to the true fishes again and firstly look at a model of fish migration that applies to many continental shelf fish (Figure 9.7A). Mature females and males aggregate at specific sites on the sea bottom—the spawning areas—and liberate eggs and sperm. The eggs either float in the plankton or lie on the sea bottom, depending on the species, and hatch to larvae which are planktonic. These latter drift in the surface current to a shallow water nursery area where they grow. The young fish then drift and swim with the current to a deeper water adult feeding area. Adults migrate from the adult feeding area to the spawning area against the current, and back again with the current. This scheme applies to a range of fish, including the eel, salmon, plaice and herring. In the North Sea the migrations of plaice

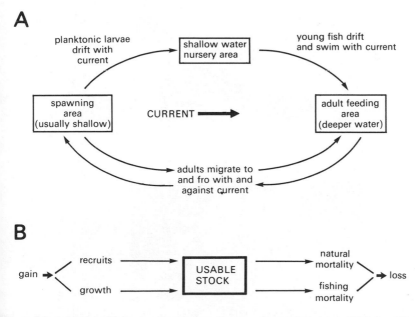

Figure 9.7 *A.* Model of fish migration that applies to many continental shelf fish (modified from Cushing, 1968).
B. Dynamics of a commercial fishery (modified from Ricker, 1958).

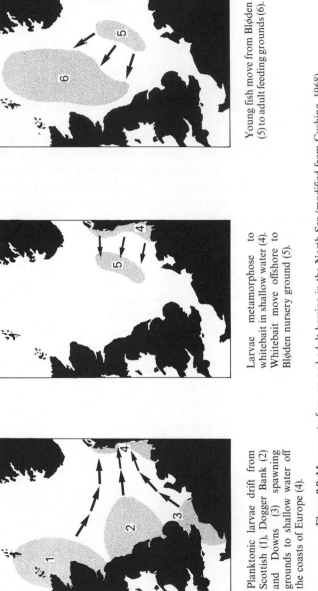

A

Planktonic larvae drift from Scottish (1), Dogger Bank (2) and Downs (3) spawning grounds to shallow water off the coasts of Europe (4).

B

Larvae metamorphose to whitebait in shallow water (4). Whitebait move offshore to Blǿden nursery ground (5).

C

Young fish move from Blǿden (5) to adult feeding grounds (6).

Figure 9.8 Movement of young and adult herring in the North Sea (modified from Cushing, 1968).

and herring are difficult to follow because, although their spawning grounds are fixed and well known, the movements of different populations, or stocks, of fish overlap. The three groups of North Sea herring provide an example (Figure 9.8). They each have separate well-defined spawning areas; the Scottish group spawn off north-east Scotland, the Dogger Bank group off the east coast on the Dogger Bank, England, and the Banks group in the English Channel. Spawning times on these areas are very precise. Larvae then drift from them to the Dutch, German and Danish coasts, where some mixing of stocks occurs. After metamorphosis the young fish or whitebait move to the nursery area, the Bløden ground. Whitebait are intensively fished by Denmark on the Bløden ground. The young fish then join the adult population (recruitment) on the feeding grounds in the northern North Sea (Figure 9.8C), where they are fished by all the nearby countries. Adult herring migrate round the North Sea in a clockwise gyre (not shown in Figure 9.8). The three groups may be partly separated during parts of this gyre, but their movements are complex and not fully understood.

Fish population dynamics

The adult stock or population of a fish species (usable stock, Figure 9.7B) depends on the balance between gains from young fish joining the population (recruits) and growth of adults in the population, on the one hand, and on losses from natural death (natural mortality) and capture of fish by fishing gear (fishing mortality) on the other (Figure 9.7B). The theory of fish population dynamics is mathematically advanced. The simplest approach is a mathematical description of Figure 9.7B for one year, thus:

$$\begin{bmatrix} \text{stock at end} \\ \text{of year} \end{bmatrix} = \begin{bmatrix} \text{stock at beginning} \\ \text{of year} \end{bmatrix} + \begin{matrix} \text{growth during year} + \text{recruitment during} \\ \text{year} \end{matrix}$$

$$- \begin{bmatrix} \text{fishing mortality} \\ \text{during year} \end{bmatrix} + \begin{matrix} \text{natural mortality} \\ \text{during year} \end{matrix}$$

The variables in this equation are measured as follows. Most fish can be aged annual rings in the otolith stones from their ears, or by annual rings in their scales. The growth rate of a population for different sizes of fish can therefore be obtained by plotting size against age. Total mortality can be calculated from the smaller numbers of larger fish caught when compared with the larger number of smaller fish caught at the same time. Natural mortality can be separated from fishing mortality by three methods.

The first relates total mortality to fishing intensity, because if fishing intensity (i.e. fishing mortality) increases, the total mortality will increase, even though the natural mortality remains the same.

The second method uses tagging data. If 100 fish are tagged and 30 are recovered by fishing, the fishing mortality is said to be 30%. However, errors are caused by fish dying after having been tagged, or by fishermen not reporting recaptures.

The third method relies on independent estimates of natural mortality. For example, since there was little fishing during World War II, mortality during that period was largely caused by natural events (natural mortality). So in the years immediately following the war, the smaller numbers of larger fish caught, compared to the larger number of smaller fish caught at the same time, was a direct estimate of natural mortality. Finally, the size of the usable stock can be measured by an analysis of landings at fishing ports, and the recruits to the stock can be measured by the numbers of the smallest and youngest fish caught each year.

Overfishing

The overfishing problem has been with mankind in one way or another for hundreds of years (Table 9.1, p. 184). A stock is being overfished when fewer fish are being caught as the fishing effort goes up (more fishing boats or better gear), and when the fish that are caught are smaller. This process is a positive feedback cycle. As fewer fish are caught, fishing effort goes up, which results in fewer fish being caught, and fishing effort goes up again. Clearly a limit to the amount and size of fish caught is desirable. The regulation of a fishery aims to produce a maximum catch with the minimum effort of fishing.

Consider the stocks of plaice in the North Sea (Figure 9.9A, B) (Beverton and Holt 1957). As fishing mortality rate (fishing effort, number of boats fishing) goes up, the yield of fish (yield/recruit, amount of fish caught) rises to a peak, but then falls again when the fishing mortality rate reaches 0.25–0.30 (Figure 9.9A). This fall is called *overfishing*. The pre-World War II fishing mortality (F) of plaice stocks in the North Sea was 0.73. The mathematical derivation of F and its units are complex (cf. Graham, 1956, p. 393). F can take any positive value from 0 to $+\infty$. If $F = 1$, the number of fish caught during the year = the size of the population at any one moment during the year, assuming that the population is continuously replenished and remains constant in size. To catch all the population in the year without replenishment would need a value of F approaching infinity.

It is possible to reduce the effects of overfishing by moving to the left up

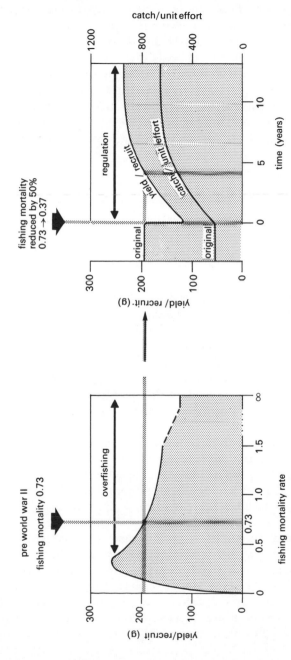

Figure 9.9 A. *Overfishing curve.* As fishing mortality rate (equivalent to fishing effort, number of boats fishing, etc.) increases, the yield/recruit (equivalent to the amount of fish caught) increases at first, but then begins to fall when the fishing mortality reaches 0.25–0.30. This fall is called *overfishing.* The pre-World War II fishing mortality of plaice stocks in the North Sea was 0.73. The mathematical derivation of the expression for F (fishing mortality rate) and its units is complex (cf. Graham 1956, p. 393). F can take any positive value from 0 to $+\infty$. If $F = 1$, the number of fish caught during the year = the size of the population at any one moment during the year assuming the population was continuously replenished (kept at a constant number of fish). To catch all the population in the year without replenishment would require a value of F approaching infinity. B. *Fishery regulation.* The predicted effect of reducing the pre-World War II plaice fishing mortality of 0.73 by 50% to 0.37. Note that the abscissa (x axis) is time in years after the 50% fishing mortality reduction. Following the reduction, the catch/unit effort (equivalent to the catch per boat) would immediately increase (lower curve and right hand ordinate). The yield/recruit, however, (equivalent to the total amount caught by the reduced fishing fleet) would take almost 5 years to recover to its previous value (upper curve and left hand ordinate) (modified from Beverton and Holt 1957, and Graham, 1956).

the curve in Figure 9.9*A*. This can be achieved by various fisheries regulation methods (see p. 183). As an example, supposing the pre-World War II fishing mortality of plaice was reduced 50% from 0.73 to 0.37 (Figure 9.9*B*) by halving the number of boats, then the catch of those boats still fishing would immediately go up (Figure 9.9*B*, lower curve and right hand ordinate). However, the total amount caught by the reduced fishing fleet (Figure 9.9*B*, upper curve and left-hand ordinate) would immediately halve, and would take five years to reach its former value. It takes this time for new fish to replace all the fish in the usable stock of the original population.

Farming the sea—world distribution

Farming the sea, sometimes called aquaculture, involves culturing marine plants and animals for food, and has been practised for hundreds of years in South-East Asia. In China its origins can be traced back 4000 years b.p. to the culture of freshwater fish. In recent years the increasing shortage of food throughout the world has led to an increase in fish farming in both developed and underdeveloped countries, and to the introduction of more efficient farming techniques. World production is about 67% fish and 33% molluscs and seaweeds. Only 25 to 33% of the fish are marine species. The rest are brackish-water or freshwater species.

   The most important commercial farming operations usually concentrate on high value species because maintenance and equipment costs are high. Cost-effective procedures are common for the same reason. These include using treated sewage as food, using waste water, and breeding hardy strains. There is also considerable interest in ranching methods. These involve artificial fertilisation and growth of young fish which are then released at sea. Japan has been practising the ranching of salmon for some years. In developing countries with coastlines, aquaculture helps rural development and employment, and provides high-quality food. The FAO and UN instituted a 10-year aquaculture development programme in the 1970s aimed at helping African, Asian, and South American countries.

   The farming of marine species is in principle more efficient than conventional fishing for either fish or invertebrates (oysters, lobsters, crabs). It is relatively easy to stabilise supply to the market, and therefore expensive machinery is not left idle for any length of time, and men can obtain continuous employment. Farmed species can also be delivered to the consumer in a fresher condition and often have a better quality than species caught at sea. For example, oysters and clams grown on farms in Europe and America fetch a better price and taste better than those caught at sea. The farming of a range of marine species now occurs throughout the world. In Europe, France has the largest industry for culturing mussels and oysters, closely followed by Spain. There are a number of centres along its Atlantic coast such as Cherbourg, St Malo, La Rochelle and Arcachon, and along its Mediterranean coast, such as Port Vendres and Toulon. These contribute to the total annual harvest of about 100 000 tonnes of artificially bred oysters and about 70 000 tonnes of cultivated and natural mussels. Oysters settle naturally on stacks of tiles placed in the intertidal region, and are grown from that stage. In Holland, the Waddensee provides a suitable area for the mass cultivation of mussels, although as yet only 5% of the total area is used. It is shallow, and is protected from the North Sea by a row of islands. Iversen (1972) quotes India as commercially farming nine species of shrimp and five species of fish, Korea as farming oysters, cockles and mussels, and Japan as farming shrimps, crabs, lobsters, fish, octopus, squid and seaweeds.

There are interesting comparisons between aquaculture in Western countries and South-East Asian countries (Table 9.2). South-East Asian countries produce and consume much more than Western countries. There are also differences in the relative proportions of categories farmed. South East Asian countries produce large quantities of seaweed but fewer molluscs.

Table 9.2 Aquaculture production in representative Western and South East Asian countries. Finfish include freshwater and marine species. (1 tonne = 1000 kg).

		Production (tonnes yr^{-1})			Consumption (g person^{-1} yr^{-1})
Western countries	*Crustaceans*	*Finfish*	*Molluscs*	*Seaweeds*	
Britain	0	5	0	0	0.1
France	0.003	25	173	0	3.74
USA	5.6	56	74	0	0.65
%	1.7	25	73	0	
South-East Asian countries					
China	0	813	1758	1441	5.57
Japan	2.5	249	298	426	8.87
Philippines	0.9	152	0.3	133	6.81
%	0.06	23	39	38	

Fish farming problems

There is a range of problems that are common to the farming of all marine species. Predators and parasites have to be controlled or eradicated, food input needs careful control, removal of waste products, faeces and so on has to be effective, and the supply of sea water, whether by natural tides or by pumps, has to be monitored for pollution. Pollution becomes especially important when molluscs are being farmed in inshore waters. The water itself is likely to be polluted by man, and the molluscs filter feed, thus picking up poisons that may kill them or pathogenic microorganisms that may kill their predator—man. Further difficulties occur when attempts are made to breed marine species in captivity, and because of this much sea farming depends on catching young animals and then growing them under controlled conditions.

The milk fish culture in Indonesia

The milk fish (*Chanos chanos*) is a herring-like species well suited to culture in brackish water ponds. It is euryhaline, grows quickly, is fairly disease-resistant, and feeds on algae. The species is common in the tropical and near-tropical regions of the Indian and Pacific oceans, from the east coast of Africa and the Red Sea, through the waters around Vietnam and the Indonesian and Australian region, to the western coast of North America. In spite of its widespread distribution and of its suitability for farming, it is only cultured on a large scale in Indonesia, the Philippines and Taiwan. Clearly the species could produce a valuable source of protein in areas such as Bangladesh, Cambodia, East Africa and South China where swamp land or similar shore line is readily available.

Milk fish spawn once or twice a year in about 25 m depth near the shore. After fertilisation, the pelagic eggs hatch in 24 hours to planktonic larvae that live on phytoplankton near the shore. Within 12 months the young are 20 cm long and about 200 g in weight, and move offshore. They become sexually mature after 6 years, and may reach a weight of 20 kg. It is not possible at present to obtain sexually mature fish under farmed conditions where the maximum length is only about 1 m.

In Indonesia, the post-larvae and juveniles are netted inshore at 15–20 mm length by commercial dealers who then sell them to the owners of the culture ponds or Tambaks. During transit, the larvae are usually fed on lightly roasted rice flour, and are slowly acclimatised to low salinities which also helps to kill off many competitors. The juveniles are placed into fry ponds and when larger they are transferred to production ponds. The ponds are dug just above high tide, so that they can be drained easily, and are about 10 cm deep for the fry and about 30 cm deep for the older fish. Their area ranges from 5×10^3 to 30×10^3 m^2. The shallow water allows a rich growth of algae to develop on the bottom, and the fish feed on this and on decaying mangrove leaves and twigs (*Avicennia*) that are added as fertiliser. Recently the red alga *Gracilaria confervoides* has been used as an additional food source. The ponds are often sited near or on mangrove swamps as the tidal flow is suitable, the bottom sediment has the correct consistency and permeability, and of course, the mangrove trees are a source of fertiliser. The most important pond pests are the larvae of the freshwater midge *Tendipes longilobus* and the mangrove snail *Cerithidea*, both of which compete with the milk fish for its food, and the marine polychaete *Eunice* whose burrows make the bottom sediment of the ponds too permeable.

Seaweed culture, *Porphyra* culture in Japan

Seaweeds are commercially farmed in Japan and China on a large scale (Bardach, Ryther and McLarney, 1972). A number of species of the red alga *Porphyra* are cultured and eaten in small quantities as a condiment, as are the green algae *Enteromorpha* and *Monostroma*, while the brown algae *Undaria pinnatifida* and *Laminaria japonica* are eaten as fresh or pickled vegetables. *Laminaria japonica* is also a valuable commercial source of iodine, bromine, potassium chloride, algin, mannitol, laminarin, and chlorophyll.

Porphyra, called laver in Europe and America and nori in Japan, is farmed in Japan as follows. In spring, chopped thalli are added to sea water in indoor tanks containing oyster shells. The thalli release spores (α-spores or carpospores) which then burrow into the shells forming filamentous growths. This *Conchocelis* stage grows as a dark red incrustation through the summer and autumn, and then releases monospores that are capable of settling on nets. At this point, nets are immersed in the tanks for about an hour and the monospores settle, after which the nets are returned to the sea. The monospores divide and grow, forming thalli which grow over winter and are harvested as food.

Culture of the *Conchocelis* stage is easy and needs little space; a tank 2.4 m $\times$ 1.8 m $\times$ 0.9 m deep will provide enough monospores for settlement on about 100 nets, and about 70% of Japan's nori production is by this method. Older methods depended on the natural settlement of monospores

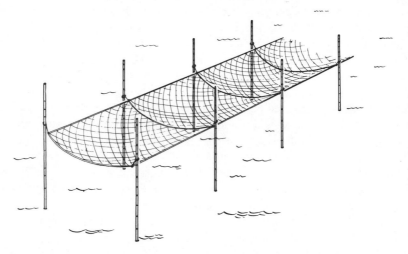

Figure 9.10 Inshore nets slung from bamboo poles which are used by the Japanese for growing the seaweed *Porphyra*. Groups of these nets form characteristic arrays in the shallow inshore waters around Japan. The seaweed hangs down into the water and is harvested at fortnightly intervals (modified from Bardach, Ryther and McLarney, 1972).

on nets in the sea, and this settlement was rather unpredictable. Present day nets measure 1.2 m wide and 18 to 45 m long, and are hung between bamboo poles so that they are parallel to the bottom (Figure 9.10). Large groups of the nets form very characteristic arrays in shallow inshore waters. The thalli are harvested throughout winter and spring, and are cut every 15 days or so, when they are about 12 cm long. They are washed in fresh water, cut into small pieces and dried into sheets — the whole process often being mechanically operated. The resultant sheets of *hoshinori* are expensive, and so the seaweed is mainly eaten as a condiment, although its protein and vitamin content are high. The *per capita* consumption in Japan is about 50 g per year, which is really very low.

CHAPTER TEN

TROPICAL INSHORE ENVIRONMENTS

The tropics contain a number of unusual and interesting inshore environments that are not found elsewhere on marine coastlines. Coral reefs are one of the great wonders of the world, and the whole structure of a coral island or atoll is dependent upon the continued growth of living organisms—the corals. They also afford shelter and a home to many populations of *Homo sapiens* in the Pacific region. Mangrove swamps, in contrast, are inhospitable. They can contain mosquitoes, crocodiles and venomous snakes, but are nevertheless an important tropical intertidal environment that should be conserved. Hypersaline lagoons in the tropics and semitropics often have extremely high salinities, and the animals and plants living in them are adapted to these. They can also contain commercially exploitable populations of fish.

Hypersaline lagoons. The Sivash

Hypersaline lagoons are semi-enclosed areas of shallow water connected to the sea, where evaporation is greater than freshwater inflow, and where salinities are above about 40‰ (Hedgpeth, 1957). They occur mainly on seashores in the sub-tropics and tropics. The best known examples are the Bitter Lakes along the Suez Canal in Egypt (31° N), the Laguna Madre in Texas (27° N) and the Sivash on the western side of the sea of Azov (46° N).

The Sivash is 1–3 m deep, and is highly productive (Figure 10.1). It has high summer temperatures and salinities which often cause mass mortality, while it sometimes freezes in winter. Young and adult Azov Sea fish continually migrate in from the sea but are limited by the hypersaline conditions. Planktonic larvae of benthic forms are carried in by currents, but are soon killed by the high salinities. In general, the Azov Sea fauna and the hypersaline Sivash fauna overlap near the mouth in salinities of 36–75‰.

In most hypersaline lagoons, the animals and plants are carried in by

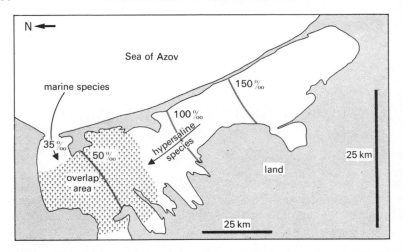

Figure 10.1 The Sivash, a hypersaline lagoon at the edge of the Sea of Azov, USSR. Marine species and hypersaline species meet in an area of overlap near the entrance to the lagoon. Salinites are given in ‰ (S. W. = 35‰). The vertical scale is twice that of the horizontal scale (modified from Emery, in Hedgpeth, 1957).

currents or migrate in from fresh water or the sea; they do not breed in the lagoons. Examples are chironomids from fresh water, gammarids from estuarine conditions, and fish, copepods, and molluscs from the sea.

Mangrove swamps, trees, geographical variation

Mangrove swamps occur in the tropics or sub-tropics between MTL and EHWS tidal levels in estuaries or on shores sheltered by coral reefs or offshore islands (Figure 10.2). They are common on some shorelines in East Africa, India, the East Indies, Australia (Queensland) and the West Indies, and can be regarded as the tropical analogue of the temperate climate salt marsh (Chapter 8). The overall structure of the mangrove swamp is dominated by a number of species of mangrove tree which vary geographically (MacNae, 1968). The Red Mangrove, *Rhizophora* species, with its long rhizophores or prop roots that reduce tidal flow, assists the deposition of sediment, and often forms almost impenetrable thickets. Most species of *Rhizophora* have shallow roots and are well adapted to grow in poorly consolidated muds. Their seeds sprout on the tree and only drop when the growing roots reach a length of some centimetres. The Black Mangrove, *Avicennia* species, is more tree-like because it has no prop roots, and

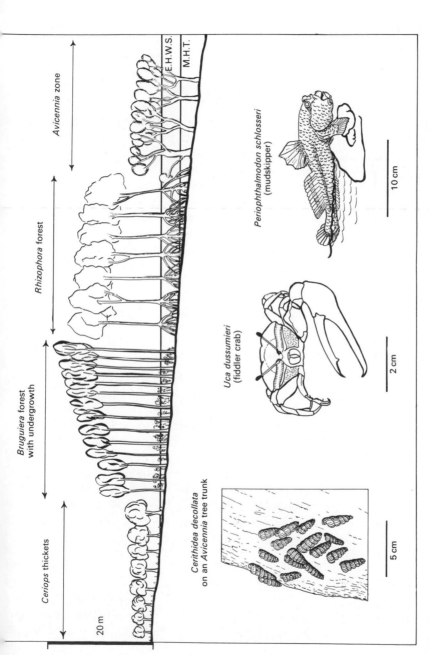

Figure 10.2 Transect across a representative mangrove swamp from above the level of extreme high water spring tides (left) to below mean high tide level (right). The numbers and kinds of trees and their heights vary geographically. Three common mangrove animals are shown in the lower diagrams. They are the snail *Cerithidea decollata* (Mollusca), the crab *Uca dussumieri* (Crustacea), and the mud skipper *Periophthalmodon schlosseri* (Teleostei) (modified from MacNae, 1968).

develops aerial roots (pneumatophores) which emerge from the ground a little way from the tree. Both *Rhizophora* and *Avicennia* can develop to full-sized trees 30 or 40 m high. *Ceriops*, on the other hand, is usually shrub-sized (1–6 m). Mangrove trees occur in zones in mangrove swamps (Figure 10.2). The landward fringe is variable, but is usually forest, grassland scrub, or desert. *Ceriops* thickets often occur towards EHWS, followed by a *Bruguiera* forest (another mangrove tree genus), then by the *Rhizophora* forest with its characteristic intertwined prop roots, and finally by a seaward *Avicennia* zone. The roots of the mangrove trees at the seaward edge are often covered by a dense algal covering. This zonation varies geographically. For instance, in Mozambique, *Avicennia* species are found near the landward fringe. Similar geographical variations occur amongst the swamp animals. In the Indo-west-Pacific region, for example, there appear to be at least eight distinct faunal and floral regions. Their centre is in the Malayan archipelago at the tip of the Malayan peninsula and south west Borneo, and from there they extend westwards to Sri Lanka and East Africa and eastwards to the Great Barrier Reef and New Zealand. The fauna becomes impoverished as one moves away from the centre, and is only partially replaced by local endemic species (MacNae, 1968).

Adaptations of mangrove trees

Mangrove trees are physiologically adapted to grow in high salinities and unconsolidated anaerobic mud (Scholander, 1968; Scholander *et al.*, 1955). Some species (*Avicennia, Aegiceras*) have salt-secreting glands in their leaves, and all have a high osmotic pressure in the leaf cell sap which allows them to take up water from seawater. In addition, their transpiration rates are low, so water loss is reduced. The prop roots of *Rhizophora* and aerial roots (pneumatophores) of *Avicennia* aerat the plant roots in the mud (Figure 10.3). When the lenticels of either species are blocked with grease, the oxygen content of the roots in the mud drops from 10–15% to 1–2%. In *Avicennia*, the oxygen content of the aerial roots is low when they are covered by the tide and high when they are exposed.

Mangrove swamp productivity

Net production of mangrove swamps in Florida, USA, is 350–500 g C m^{-2} yr^{-1} which is considerably higher than the 50–75 g C m^{-2} yr^{-1} of the phytoplankton in the surrounding waters. Maxi-

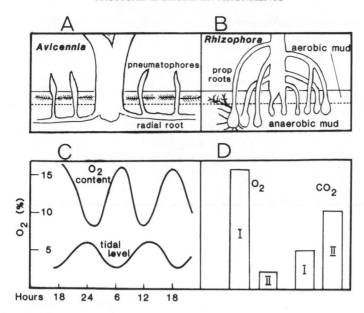

Figure 10.3 Adaptations of mangrove swamps, trees and root systems (after Scholander, van Dam and Scholander, 1955; Mann 1982). *A*: genus *Avicennia* the black mangrove, and *B*: genus *Rhizophora* the red mangrove. *C*: reduction in O_2 content of radial roots of *Avicennia* when covered by the tide. *D*: reduction in O_2 content and increase in CO_2 content in the prop roots of *Rhizophora* when the lenticels on the roots are covered with grease to prevent gas exchange. I, normal (control); II grease-covered.

mum production occurs in the middle of the tree canopy rather than at the top, particularly on sunny days because water stress on the leaves at the top of the canopy causes stomatal closure.

About 5% of the leaf production of red mangroves in Florida is eaten by terrestrial grazers (secondary production) (Odum and Heald, 1975). The remaining 95% enters the intertidal or estuarine environment as debris. In the North River system in Florida, about 40 out of 120 species of animals take in 20% or more of their diet as vascular plant detritus. The 40 species include fish, crustaceans, polychaetes and insect larvae. These detritivores can be grouped as grinders that chew large leaf particles, deposit feeders that eat small leaf particles on the sediment surface, and filter feeders that ingest fine particles in suspension. Clearly mangrove leaf litter is a major source of food and energy in this very productive system (Mann, 1982).

Mangrove swamp sediments

The mangrove swamp sediment has a low oxygen content and high hydrogen sulphide content. It is often semi-fluid or contains mangrove peat made up of decaying roots and branches. Calcareous debris from foraminiferan and mollusc shells is also common and is brought in by the tide. The aerial roots of *Rhizophora* may have many sedentary marine invertebrates attached to them (oysters, tunicates and sponges). Towards high tide, sediment may be mined by the mud lobster *Thalassina anomala* which produces mounds up to 3 m high. The salinity of the water in the sediments ranges from almost fresh water to 100‰, while the swamp itself usually has tidal run-off channels. These channels can, in fact, be used to explore the swamps by small boat.

Mangrove swamp fauna

The animals in mangrove swamps invade it from the surrounding land, fresh water, and sea. The synchronous flashing of fireflies throughout the night on mangrove trees is dramatic, while birds, mammals and a number of snakes are fairly common. For example, in some Australian swamps large aggregations of flying foxes (*Pteropus*) occur. Intertidal *Littorina* species are zoned horizontally and also vertically up the mangrove trees, the robber crab *Birgus latro* inhabits the landward fringes, and at high tide level the burrows of the crab *Cardisoma carnifex* are common. Other species of crabs of the genus *Sesarma* live in *Thalassina* burrows in the landward fringe, and in the *Ceriops* thickets and *Bruguiera* forests. Both these latter contain large numbers of molluscs (e.g. *Cerithidea*), and mobile crustaceans (e.g. *Uca* species). Moving seaward, we find that the *Rhizophora* forest fauna is very similar, and their prop roots are colonised by many sedentary marine invertebrates (tunicates, bivalves, and barnacles). the *Avicennia* zone contains more truly marine animals. The mudskipper *Periophthalmodon* occurs in this zone, and has a characteristic pattern of behaviour by which it defends its bowl-shaped burrows against intruders.

Coral reefs

Coral reefs occur around the world in warm sub-tropical or tropical oceans where the mean annual temperature is about 23 to 25 °C (Figure 10.4). For full development they need ample sunlight, as corals do not grow well below 20–30 m. Their rate of calcification decreases with increasing depth, and

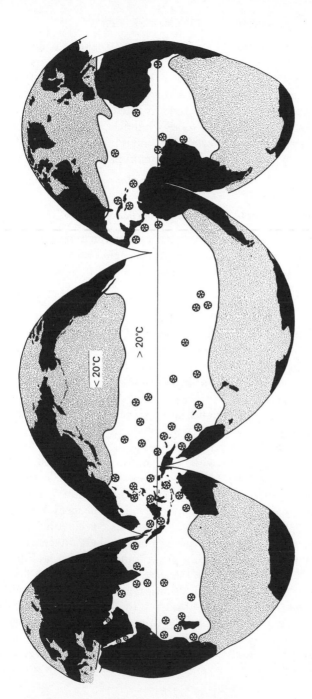

Figure 10.4 Global distribution of major coral reef areas (enclosed stars). Waters colder than about 20 °C are shaded. Coral reefs are restricted to tropical oceans where the mean annual temperature is about 23 to 25 °C (modified from Hedgpeth, 1957).

species that are massive near the surface are only weakly calcified at depth. They also need a good circulation of sea water bringing oxygen, carbon dioxide and nutrients. Their framework consists of hermatypic reef-building corals and calcareous algae with sediments formed from their dead remains; all of these are interlocked and cemented together by coral growth to form a solid structure with buttresses, channels, overhangs and tunnels. Many sedentary and mobile invertebrates and brightly coloured fish live on or in reefs. As a whole, therefore, the community is extremely rich and varied.

Factors limiting coral reefs, growth and reproduction

The main factors limiting the development, maintenance and diversity of coral reefs are geographical area, temperature, light, water depth, turbidity salinity, and biotic factors such as predation by fish and *Acanthaster* (Levinton, 1982).

The geographical area in which a coral reef is found has a major effect on the whole coral reef ecosystem: Indo-Pacific coral reefs are very different from Atlantic coral reefs. In general, coral reefs do not occur below about 20 °C, although reefs in the Florida Keys, USA grow at 18 °C. At the edge of their distribution episodic low temperatures can cause mass mortalities. For instance in 1968 air temperatures in the Persian Gulf fell to 0 °C and water temperatures to 10 °C. As a result, almost all the inshore coral reefs were destroyed. On the other hand, hermatypic corals can grow at 14.5 °C off South California, although they do not form reefs.

Light is very important because zooxanthellae need it to photosynthe sise. It decreases exponentially with depth and so has a controlling effect on the deeper parts of a reef. Reef building is reduced below 25 m in the Indo-Pacific, and in the Caribbean there is little growth below 75 m. However those corals adapted to the lower light intensities at deeper levels of the ree often die if transplanted to higher light intensities. Species diversity usuall decreases with depth, but the reverse can occur where adverse condition affect species at the surface. For example near the sea surface at Eilat, Israe adverse temperatures, salinity, and desiccation inhibit species diversity.

Corals cannot grow in lowered salinities; however, they can withstand higher salinities of, for example, 40‰ as in the Persian Gulf. Even heav rains or run-off from rivers during the rainy season can kill a significan number of colonies; this is common on the north coast of Jamaica.

High turbidity and sedimentation inhibit coral growth by reducing th light intensity and also by covering the colony surface with silt. This ca have a marked local effect. The northern windward coast of Jamaica has clear waters and good coral growth, while the waters of the southern leeward coast are more turbid, and coral reefs are less well developed there. Corals remove silt by producing mucus which is carried across the colony surface. It is interesting that more mucus is produced by massive slow-growing forms such as *Platygyra* than by thin branched species such as *Acropora palmata*. Those species most adapted to lagoons or estuaries are resistant to silting and burial, as well as to low salinities and varying temperatures (Levinton, 1982).

Coral reefs can be badly damaged by hurricanes, and the occasional catastrophic effects in Jamaica, West Indies, and Queensland, Australia, are well documented. Huge blocks of coral are torn up and whole reefs broken apart. However this can also transport pieces of living coral to other sites and produce new bare surfaces for colonisation by coral larvae (planulae). Reefs are therefore dynamic ecosystems that are constantly changing, sometimes very suddenly as a result of environmental damage.

A number of endolithic organisms bore into coral reefs. These include sponges, polychaetes, molluscs and sipunculids. Many reef fish also eat coral heads. This bioerosion weakens the fabric of the reef. For example clionid sponges often live in the base of coral skeletons (Goreau and Hartman, 1963) and may weaken the base so that it eventually breaks. The weakening is probably caused by the sponges chemically dissolving the calcium carbonate and by sponge contraction loosening small calcium carbonate chips. These chips may be an important part of the fine silt fraction of reef sediments. On some reefs in Jamaica bioerosion is a major factor controlling reef morphology, especially at depth, where the large spherical forms fall down the slope when they break.

Geographical variation in reefs, Indo-Pacific and Atlantic Provinces

Coral reefs can be divided biogeographically into the Atlantic Province and the Indo-Pacific Province reefs. The reefs in the Indo-Pacific differ from those in the Atlantic by being grouped into rings or chains of islands or atolls, by containing many more coral species, and by having much richer populations or corals on their intertidal reef flats.

In the tropical Atlantic (Caribbean, Bahamas, Brazil, Gulf of Guinea, West Africa) there are fewer coral species than in the Indo-Pacific, about 35 compared with about 700. However *Acropora* and *Porites* are the dominant genera in both areas. Calcareous algae play a very minor part in the construction of the Atlantic reefs, while many invertebrate species are absent (the giant *Tridacna*, the diverse mollusc fauna, coral gall crabs, giant sea anemones with their commensal fish and crustaceans).

In the Indian and Pacific oceans the coral reef species are remarkably constant, from the Red Sea and east coast of Africa, through the Indian Ocean and the Great Barrier Reef, Australia, to the Tuamotu Archipelago in the Pacific (NE of New Zealand, 15° S, 145° W), a distance of about 22 500 km. However, as one moves north and south outside the 18 °C isotherm, the numbers of species, genera, and families all fall rapidly. For example, from the northern end of the Great Barrier Reef in Torres Strait (9° S) for about 1440 km to 21° S (about 300 km SE of Townsville), 60 hermatypic coral genera occur. Within the next 400 km to about Rockhampton the number of genera falls to 20 as the winter sea temperature drops below 18 °C.

Hermatypic corals, role of zooxanthellae

Coral reefs are mainly built by hermatypic corals. These corals are scleractinian coelenterates which deposit a heavy $CaCO_3$ skeleton and contain zooxanthellae. Calcium and carbonate are obtained from food and by absorption from sea water. The corals live in colonies with the living tissue as a thin layer over the skeleton. In most species the polyps only open at night when the reef plankton is most abundant, and feeding occurs then.

Small symbiotic photosynthetic algae, zooxanthellae, live in endoderm cells of living coral, and soluble metabolic by-products of photosynthesis pass from them to the coral tissue. Zooxanthellae have a number of roles in coral metabolism. They play a major part in reef building by causing faster skeleton formation (calcification) during daylight hours. The mechanism is not clear, however. Zooxanthellae use phosphate during photosynthesis. Phosphate also inhibits the crystallisation of aragonite—the form of calcium carbonate in coral skeletons. The removal of phosphate by zooxanthellae may therefore allow the rate of calcium carbonate formation to increase. Zooxanthellae might also produce proteins that help the formation of the organic matrix within which calcium carbonate is laid down by the coral.

Zooxanthellae aid in lipogenesis by hermatypic corals. Lipid makes up about 33% of the dry weight of anemones and corals, and is probably an important energy source. In some corals lipogenesis can increase by up to 300% in some corals during daylight, and this implicates zooxanthellae. Fatty acids taken up by corals as they eat plankton are oxidised to acetate in the digestive cells, and the zooxanthellae are thought to convert the acetate to saturated fatty acids which eventually form lipids. The detailed biochemistry is not fully understood.

Zooxanthellae may be directly utilised by corals as a food source, or may release photosynthetic by-products such as glycerol, glucose and alanine to the coral tissue. Interestingly enough, corals can also take up glucose directly from sea water.

When measured as plant and animal protein, some corals contain as much plant tissue (zooxanthellae) as animals tissue (coral), while some corals can live and grow for months by using zooxanthellae metabolic products when no other food source is available. On average, however, a coral contains three times as much plant as animal tissue, but only 6% of the plant material is zooxanthellae; the remainder is made up of filamentous green algae in the skeleton.

Small zooxanthellae occur in other reef organisms, such as colonial tunicates (*Diplosoma*), foraminiferans (*Polytrema*), the coralline hydroid *Millepora* and sponges. Zooxanthellae are also found in the mantle edge of the giant clam *Tridacna*. This mollusc, which may be a metre long, plays a major part in the economy of the Great Barrier Reef, Australia, but unlike the corals actually digests its zooxanthellae.

Coral reef productivity

Reef productivity is very high, $300-5000 \, g \, C \, m^{-2} \, yr^{-1}$, when compared to the productivity of surrounding waters, $20-40 \, g \, C \, m^{-2} \, yr^{-1}$. The reasons for this marked difference are still not fully understood (Lewis, 1977). The concentration of nitrogen and phosphorus in waters flowing over reefs is consistently low, and these nutrients are probably recycled within the reef system by bacteria and the primary producers. The main primary producers are zooxanthellae, marine grasses, calcareous algae, fleshy macrophytes, and filamentous algae on corals or rock.

Detritus is probably more important to the coral reef ecosystem than is zooplankton. The mass of detritus in the water above the reef is greater than the mass of zooplankton and is probably a more important food for corals and other benthic organisms than is zooplankton. Bacteria may also be important in nitrogen fixation, decomposition and biogeochemical cycling. The quantitative significance of grazing and predation by the abundant reef fish fauna has not been fully analysed.

Energy budgets of individual corals show that they produce a large excess of photosynthetically fixed energy (Edmunds and Davies, 1986). *Porites porites* on the fore-reef at Discovery bay are largely autotrophic, making little use of zooplankton as a food source. Of the daily photosynthetically fixed energy, 33% is used for animal respiration and growth, 22% for zooxanthellae respiration and growth and less than 1% for reproduction as planulae. This leaves a remarkable 45% that is not accounted for.

Interspecific competition in corals

Interspecific competition for space is important in environments where large numbers of organisms are crowded together, as on a coral reef, an intertidal rocky shore or a mud flat. It occurs between coral species on coral reefs by three mechanisms: shading, overgrowing and eating.

Foliose and rapidly growing forms such as some species of *Agaricia* shade the species underneath them which often grow more slowly. Some species physically grow over others and thereby inhibit and eventually kill them.

Finally, some species show a most interesting phenomenon of *interspecific digestion* (Lang, 1973). A hierarchy exists on Jamaican coral reefs, in which species higher in the hierarchy eat those lower down. *Scolomya lacera* and *Mussa angulosa* are the two most dominant species of the eleven investigated by Lang, and *Scolomia cubensis* and *Isophyllastrea rigida* are the least. The dominant species extends mesenteric filaments towards subdominant species and digests those parts of subdominant species that it can reach. Interspecific digestion by *extracoelenteric feeding* may well be common on all coral reefs.

Coral reproduction and growth

Sexual reproduction in corals is usually internal. The larvae develop in the gastrovascular cavity and are released as planulae. These enter the plankton and act as a dispersal stage. They eventually settle on a hard substrate, and metamorphose to a single polyp. The polyp divides a number of times to form a true coral colony containing large numbers of polyps. The recruitment rates of different species (settlement rate) are very variable.

Branching corals grow more quickly than hemispherical ones. In the Caribbean, the staghorn coral *Acropora cervicornis* grows up to $10 \, \text{cm} \, \text{yr}^{-1}$ (branch tip extension) while the hemispherical *Montastrea annularis* grows up to $0.7 \, \text{cm} \, \text{yr}^{-1}$ in height. However, the massive hemispherical species usually reach a larger size when measured by weight. *M. annularis* has different growth forms in shallow and deep water, and this has a significant effect on coral reef topography. At water depths shallower than 10 m, the species forms large hemispherical colonies which are important in the formation of reef buttresses. In deeper water (30 m) plate-like forms grow laterally from almost vertical coral reef cliffs.

Colony growth can be measured in several ways (Levinton, 1982). These include measuring growth from defined reference points or metal spikes, the use of the dye Alizarin Red-5 as a pulsed marker incorporated into the skeleton, measurement of growth bands, and uptake rates of ^{45}Ca and ^{14}C.

Types of coral reef

Four main types of coral reef can be recognised — atolls, barrier reefs, fringing reefs, and platform reefs. Atolls occur in the Indian and Pacific Oceans (known as Indo-Pacific Atolls); barrier reefs such as the Great Barrier Reef off Queensland, Australia, and the Great Sea Reef north-west of Fiji, run parallel to and some kilometres from the coastline proper; fringing reefs develop close to islands like those on the north coast of Jamaica, and platform reefs form in the lagoons behind barrier reefs. These types are all structurally related, and may merge one into the other. Their relations become clearer when one looks at the way coral reefs form.

The South Pacific has a number of chains and rings of atolls most of which are inhabited by man. Kiribati (Gilbert Islands, Kingsmill group), lying on the equator between 173° and 177° E, and north-east of Australia, is a typical example. It consists of a chain of atolls and islands having a land area of 298 square kilometres and is inhabited by about 60 000 people. The soil is of poor quality, and the vegetation is mainly coconut and *Pandanus* trees. However, the lagoonal systems offer a wide variety of fish. Line fishing is practised on the reef flats, lines and nets are used in the lagoon itself, and also in the ocean adjacent to the atoll.

Atoll formation

Submarine volcanoes that erupt in the tropics to form islands are colonised by corals and coralline algae at their edges. Colonisation and growth is limited to shallower than about 50 m by light attenuation in the sea, and sediments soon form from coral and algal debris produced by wave action. Aerial and submarine erosion will act together to level the island, while the island itself will slowly sink by its own weight. As long as this levelling and sinking is slow enough, corals will keep pace by growing upwards, keeping within 0 to 50 m depth, and thus maintaining the reef at the sea surface (Figure 10.5A). Soon the volcanic core is surrounded by a lagoon between it

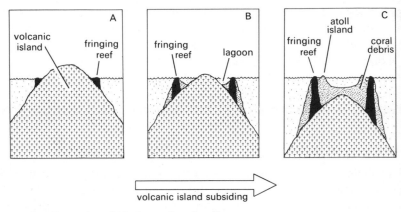

Figure 10.5 Formation of fringing reefs and atolls.
A. A newly formed volcanic island is colonized by corals to produce a fringing reef.
B. The island sinks and the corals grow so that they keep within 20 m of the sea surface. Coral debris forms sediments (light shading) and a lagoon develops between the island and the fringing reef.
C. The island sinks below the surface of the sea, but is by now covered by coral debris forming a thick layer. Islands develop in the lagoon, eventually forming a typical atoll made up of a ring of islands within the fringing reef.

and the fringing reef (Figure 10.5*B*), and eventually the volcanic core subsides below the sea, more sediments are produced by erosion, and these form a ring of islands within the fringing reef. The end product is a coral atoll (Figure 10.5*C*).

Atoll morphology

Coral atolls, such as the Maldive Islands, and the Eniwetok, Funafuti, and Bikini groups range in size from under 1 km to hundreds of kilometres in diameter, and may be circles, ovals or curved chains of islands. In their simplest form a sheltered lagoon separates windward and leeward reefs, on which are low islands formed from coral and algal debris (Figure 10.6). Prevailing winds impose an asymmetry on the atoll. They shift the islands

Figure 10.6 Transect across a representative coral atoll from the Indo-Pacific (modified from Hedgpeth, 1957).
The three species of coral illustrated in the lower part of the figure are Atlantic species, and were collected from a fringing reef on the south coast of Jamaica, West Indies. *Acropora palmata* is a shallow-water species, being usually found between 0 and 10 m. *Acropora cervicornis* is more delicate and lives at depths of 10 to 30 m. *Colpophyllia natans* is a massive species that grows at about 20 m.

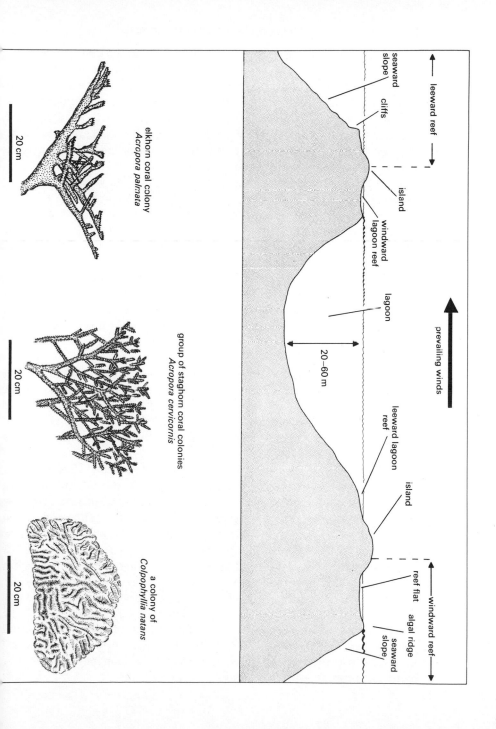

leeward reef

seaward slope

cliffs

island

windward lagoon reef

lagoon

20–60 m

prevailing winds

leeward lagoon reef

island

reef flat

algal ridge

seaward slope

windward reef

elkhorn coral colony
Acropora palmata

20 cm

group of staghorn coral colonies
Acropora cervicornis

20 cm

a colony of
Colpophyllia natans

20 cm

towards the leeward edge of both the windward and the leeward reefs and increase the width of the windward reef. They also impose a constant angle on the seaward slope of the windward reef in comparison with the cliffs and overhangs of the seaward slope of the leaward reef. If the leeward reef is far enough from the windward reef it develops a small windward reef, the windward lagoon reef, on its windward side. The reefs are broken by channels through which the sea enters and leaves the lagoon. These latter are more common on the leeward reefs.

A reddish-purple ridge is formed at the edge of the windward reef by calcareous red algae (*Porolithon, Lithothamnion, Lithophyllum*). It grows to about 50 cm above low tide in heavy surf, contains few coral species, and is broken by smooth-sided surge channels. The surge channels can extend 20 m down the seaward slope of the reef, are often almost covered over by calcareous algae and corals (*Turbinaria, Acropora, Millepora*), and carry debris and sediments off the reef flat. Buttresses of coral often form between the channels running down the reef. The seaward slope from 0 to 20 m is difficult to study because of the very heavy surf. Towards the wave base more coral species appear, some of which may be enormous. Below 20 m, corals with a more leaf-like form, such as *Echinophyllia* and *Oxypora*, can grow.

The windward reef flat deepens from a few centimetres near the algal ridges to 2 to 3 m some metres away. It is a varied habitat up to about 1000 m wide. Small micro-atolls of *Acropora* 1 to 2 m diameter and 0.5 to 1 m height, as well as many other corals, grow on the sand or gravel bottom, and many mobile and burrowing invertebrates live in or on the sediments and corals.

The leeward reef is sheltered from strong winds and heavy surf and breakers, and this is reflected in its morphology. On its seaward slope, no algal ridge develops, and anastomosing branching and leaf-like corals grow to within about 3 m of the sea surface (*Acropora, Porites, Millepora*) as well as more massive forms (*Favia, Porites*). Mobile animals are vary abundant and vertical cliffs of corals are common.

The windward lagoon reef and the leeward lagoon reef are similar to, but less well developed than, the windward and the leeward reef.

The lagoon between the reefs is usually 20 to 60 m deep. It is floored by sediment and debris from corals, foraminiferans, and calcareous algae (*Halimeda*). Many living *Halimeda* and a few thickets of the branching coral *Acropora* colonise the floor. Patch reefs develop in shallow water, and coral knolls rise from the deeper parts of the lagoon. The sediment is inhabited by many sand-dwelling invertebrates.

Fringing reef morphology

A fringing reef, like that at Discovery Bay on the north coast of Jamaica, is similar to the windward reef of an atoll, but has no algal ridge. A shallow back reef zone, 1 to 10 m deep, extends from the shore outwards to the reef flats. It is floored with patches of living coral and algae and sandy calcareous sediments of dead coral and calcareous algae (*Halimeda*), and contains many benthic invertebrates. The reef flat 200–600 m offshore is often less than 1 m deep. Many branching and spherical corals live here, as well as spiny sea urchins (*Diadema*), sponges, sea fans (Gorgoniaceae) and coral fish. At the outer edge of the reef flat, buttresses and sand channels develop like those on the seaward slope of a coral atoll, and the large branching elkhorn coral *Acropora palmata* points into the breakers. At a depth of 20–30 m, wave energy is greatly reduced, and the corals are less massive, but there are many more species. Typical species in this region are the more delicate staghorn coral *Acropora cervicornis*, and the lettuce-like *Agaricia tenuifolia*. Between 25 and 70 m, the reef slope is often nearly vertical with a variety of delicate corals and gorgonians.

Coral reef fish

Fish on coral reefs are brightly coloured with contrasting pigments which match the marked contrast of light and shade on the reef. Many are perch-like such as the butterfishes (Chaetodontidae) and damsel fishes (Pomacentridae), which feed on corals and other invertebrates. Parrot fishes (Scaridae) have beaks of fused jaw teeth and pharyngeal teeth with which they crush and then eat corals, encrusting algae and limestone debris.

Coral death. The crown of thorns starfish (*Acanthaster*)

The form and size of a coral reef is determined by the balance between coral growth, and coral death and disintegration. Corals are killed by exposure to air, by surface layers of fresh water caused by tropical rain stroms, and by heavy surf and waves. They are also eaten by many reef-dwelling organisms. The most destructive of these is probably the crown-of-thorns starfish *Acanthaster planci*, which feeds on the polyps of living coral, leaving dead coral skeletons (Figure 10.7). The population explosion of *Acanthaster* since about 1960 to date has caused widespread destruction of coral reefs in the Indo-Pacific region. In some areas its effect has been devastating, as herds of *Acanthaster* move from reef to reef across sandy bottoms at speeds of up to 3 km/month, while in others it is known as a normal reef inhabitant

H

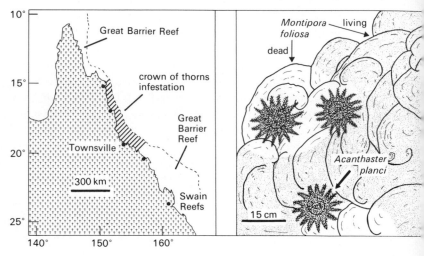

Figure 10.7 *Left*: Approximate extent of the crown of thorns infestation along the Great Barrier Reef 1972/3 (modified from Jones and Endean 1976).
Right: Crown of thorns starfish *Acanthaster planci*, feeding on the coral *Montipora foliosa* (Great Barrier Reef, Australia) (modified from Jones and Endean, 1976).

occurring at low population densities and having little deleterious effect; the explanation of these differences in population density is not known. When the corals die the normal invertebrate and fish fauna disappear. After death, the coral skeletons are soon covered by green algae, mainly *Enteromorpha*, then by calcareous algae, and after some months by alcyonarians. After several years new corals settle on the dead skeletons. Corals on *Acanthaster* infested reefs may survive if they are in positions exposed to heavy waves. The present huge population explosion of *Acanthaster planci* is probably unique although earlier minor ones are recorded. Reefs may need decades to recover from *Acanthaster* infestation.

CHAPTER ELEVEN

MAN'S USE OF THE OCEANS

This chapter describes a number of applied aspects of marine science which are central to man's use of the oceans. Maritime law is becoming a major factor in determining which nations have access to the sea and sea bed for oil and mineral exploration. Position fixing and satellite imagery have developed to a point at which they can be used to position a ship within 10 m, to detect objects less than a metre in size, and to plot coastal fronts and oceanic productivity over areas as large as the North Sea. The exploitation of oil and gas and the mining of sea-bed minerals on the continental shelf have become so important that some nations' economies depend on them. The world's demand for energy has stimulated many coastal states to examine the possibility of obtaining energy from waves. tides and ocean heat. Finally, all of those activities are potential sources of pollutants in the oceans, which cover three-quarters of the earth's surface.

Maritime jurisdiction zones — law of the sea

The first United Nations Conference on the law of the Sea (UNCLOS I) was held in Geneva in 1958, and a second (UNCLOS II) was held in 1960. Before then, maritime law was based on rules of cutomary law or treaties between states. At UNCLOS I, most of the fundamental rules of maritime law at that date were incorporated into four conventions: (1) the Territorial Sea and the Contiguous Zone; (2) the High Seas; (3) Fishing and Conservation of the Living Resources of the High Seas; (4) the Continental Shelf. The four Conventions were called the Geneva Conventions on the Law of the Sea, 1958, and are still in force. UNCLOS III met from 1973 to 1982, held 12 sessions extending over about 95 weeks, and was attended by 150 states. The New United Nations Convention on the Law of the Sea produced by UNCLOS III was signed in December 1982 in Jamaica. Unfortunately only 117 of the 150 states signed, and these did not include Italy, Japan, the United Kingdom, the United States, and West Germany. There are now seven generally accepted maritime jurisdiction zones that extend seawards from a coastal state. These are *internal waters*, the *territorial sea*, the *contiguous zone*, the *exclusive fishing zone*, the *exclusive economic zone*, *archipelagic waters*, and the *high seas*.

Almost all states (countries) with marine coast lines now claim legal jurisdiction over zones around their coasts (Figure 11.1). Bays, ports, estuaries, rivers and waters landward of the low-tide line and other semi-enclosed areas are *internal waters* over which the coastal state has complete sovereign jurisdiction, and over which all the laws of the state apply. This sovereign jurisdiction also applies to the territorial sea and contiguous zone. There is no guarantee of *innocent passage* to foreign ships in internal waters.

Innocent passage is 'innocent' as long as it is not prejudicial to the 'peace, good order or security' of the coastal state. It excludes weapons practice, fishing, research activities, and 'any other activity not having a bearing on passage'. The research activity limitation means that research vessels must obtain permission from a coastal state, and they usually have to carry an observer from the coastal state. Where innocent passage is allowed, coastal states can limit it to specified *sea lanes*, particularly for tankers and nuclear powered ships. In addition, ships in innocent passage that are carrying

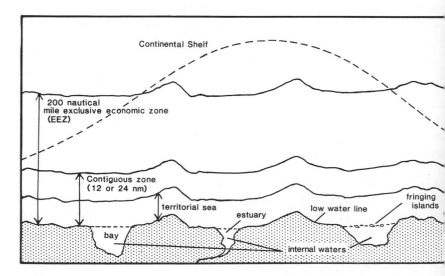

Figure 11.1 Maritime jurisdiction zones now claimed by most coastal states (after Couper, 1983).

nuclear or other dangerous substances must carry documents and observe internationally agreed precautionary measures.

The *territorial sea* extends 12 nautical miles from the coast or internal waters and is measured from low water level. Within it the coastal state guarantees a right of innocent passage to foreign ships.

The *contiguous zone* contains within it the territorial sea and extends 24 nautical miles from the coast. Foreign ships have the same rights of passage as in the EFZ (see below) or on the high seas. Where overlap between these zones occurs, the laws governing the territorial sea take precedence over those of the contiguous zone, and these take precedence over those of the exclusive fishing zone and exclusive economic zone.

The *exclusive fishing zone* (EFZ) is a legal product of the difficult period in the 1960s when there was considerable discussion between states. It extends 200 nautical miles from the coast, and the coastal state claims all rights over the fisheries within it. EFZs are being progressively replaced by exclusive economic zones which offer a more comprehensive package of legal controls to the coastal state.

The *exclusive economic zone* (EEZ) extends 200 nautical miles from the coast and in it the coastal state has sovereign rights of exploiting, conserving and managing the natural resources of the sea bed, subsoil and overlying waters. This includes fish and fisheries in the EEZ. The coastal state also has jurisdiction over man-made artificial islands and structures such as oil rigs, and controls marine research within the zone. The recent introduction of the EEZ has had considerable implications for marine research, marine exploration, and freedom of navigation. However the coastal state also has legal obligations. Within the EEZ, it must manage the living resources so that they are not over-exploited. Animal populations must be conserved to yield a *maximum sustainable yield*. The coastal state must then determine a *total allowable catch* and its own *harvesting capacity*, and if the former exceeds the latter the surplus must be made available to other states.

Islands and archipelagic states are grouped as a *regime* of *islands* in UNCLOS III. Islands are treated like mainland territories and are defined as a naturally formed area of land surrounded by water, which is above water at high tide. They have a territorial sea, contiguous zone, EEZ and continental shelf. However, rocks which cannot sustain human habitation or economic life of their own, such as Rockall, do not have an EEZ or continental shelf, but are allowed a territorial sea.

Mid-ocean archipelago states such as the Philippines or Indonesia are termed *archipelagic states*, have *archipelagic waters*, and an *archipelagic base line* which is not low tide level. the archipelagic base line is a series of straight lines joining the outermost islands and reefs of the archipelago and divides the archipelagic waters inside it from the territorial, contiguous and exclusive economic zones outside.

The *high seas* are all the remaining parts of the worlds oceans that are not included in the EEZs of coastal states, islands and archipelagic states. In spite of the introduction of the EEZ the high seas still cover most of the world's oceans. The *freedom of the high seas* includes navigation, fishing, laying submarine cables and pipelines, overflight, constructing artificial islands, and scientific research. These freedoms must be exercised by each state with due consideration for the same freedoms of other states, and for international rights in or on the ocean floor and subfloor such as sea-bed mining.

The legal definition of the *continental shelf* is the sea bed and subsoil to the edge of the continental margin (shelf + rise + slope), or to 200 nautical miles from the low-tide base line, whichever is greater. Considerable legal problems exist if the continental margin extends beyond 200 nautical miles. The coastal state has sovereign rights over the sea bed and subsoil of the legal continental shelf (mining, fishing, natural living and mineral resources) but these do not affect the legal status of the waters above the shelf. The coastal state can erect oil and gas rigs and define 500 m safety zones around them, but freedom of navigation exists. Coastal states normally grant consent for scientific research on or in the sea-bed of the continental shelf or EEZ but there may be problems where applied or commercially oriented research is proposed.

Navigation and position fixing

Navigation and position fixing in small ships within site of land is usually straightforward, often only involving landmarks, a compass and an echo sounder. Most inshore scientific work also needs a careful use of nautical charts, tide tables, a chronometer, and sometimes a sextant. It is also important to know the position of buoys, navigation buoys in estuaries, and lighthouses.

The nautical chart is probably the most important navigational and position fixing aid, particularly close to land. It is drawn on a Mercator projection and so compass bearings are straight lines.

Many nations produce their own charts, and international standardisation is co-ordinated by the International Hydrographic Organisation, Morocco. In the UK, charts are produced by the Hydrographic Department of the Ministry of Defence. Charts show depths, depth contours, underwater obstructions, Decca navigation lines, and in nearshore water, navigational channels, lights and coastal details.

Navigation and position fixing by larger ships, particularly when on a long ocean cruise is by one of three types of navigation system: inertial, hyperbolic, and satellite.

(i) *Inertial systems*. The most recent is the Ship's Inertial Navigation System, SINS. This is a self-contained dead reckoning system which monitors the ship's motion and hence plots its position. It is expensive, and is only used on military vessels.

(ii) *Hyperbolic systems*. These depend on differences in time of arrival of synchronised signals from different radio transmitters. *Decca* has a chain of transmitters whose range is about 460 km day and 190 km night. Its accuracy is 460 to 3700 m. The *Loran* system—long range navigation—has a range of 2220 km, and an accuracy of 15 to 2000 m, and the *Omega* system, whose stations have a global coverage, is accurate to 1.8–3.7 km.

Radar is another well-known position fixing device which is mainly used for inshore navigation and for avoiding collision in fog or at night. Radio signals are transmitted from an aerial on the ship and are reflected back from objects such as the coast, buoys and other ships. Its maximum range at sea is about 110 km.

(iii) *Satellite systems* (see Figures 11.3, 11.4, 11.5). The *TRANSIT* system employs five satellites in polar orbits of periods of 108 minutes. This gives position fixes accurate to 300–400 m at intervals of between 30 minutes and 3 hours. A more accurate high orbital system called *NavStar* is planned and will give a continuous position. The *International Satellite* gives ships contact for distress, safety, medical assistance, weather forecasts and social communications. It has operated since 1982.

Remote sensing

Remote sensing (RS) of the oceans has been conducted for many years, but has rapidly expanded since the advent of satellites (Lintz and Simonett, 1976; Cracknell (ed.), 1981, 1983). RS means gathering scientific information about the environment by equipment that is remote from the habitat or ecosystem being studied. In oceanography the equipment is carried on ships, aircraft or satellites.

Remote sensing systems on ships use sound waves transmitted through water to give information about animals in the water (e.g. fish schools), and about the structure of the sea bed and also for position fixing (Figure 11.2).

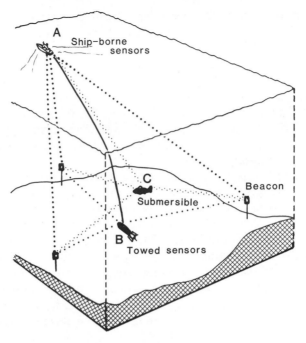

Figure 11.2 Acoustic position fixing. Uses: positioning a ship over a point on the sea bed; positioning a deep-towed instrument package relative to the sea bed. Surface ship (*A*), towed instrument package (*B*), and submersible (*C*), fix their position by timing a sound pulse to and from a trio of beacons, from the sea bed, and, for *B* and *C*, from the ship. The ship may also position itself by satellite (after Ingham, 1975).

This latter includes echo sounding and side-scan sonar of sea-bed bathymetry, sea-bed topography and wrecks, and refraction and reflection profiling of sediment strata and rocks below the sea bed (p. 235), often in relation to the oil industry.

Remote sensing systems on aircraft and statellites use electromagnetic rediation transmitted through the atmosphere which detects structure and topography at or just below the land and sea surface (land forms, surface currents, sea temperature, waves).

Remote sensing from aircraft and satellites

All objects can reflect, absorb, or emit energy from different parts of the electromagnetic spectrum (Strahler and Strahler, 1983) (Figure 11.3). It is this which forms the basis of RS from aircraft and satellites. Different pieces

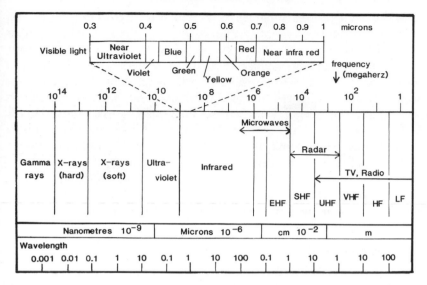

Figure 11.3 The electromagnetic spectrum (after Strahler and Strahler, 1983).

of RS equipment use the ultraviolet, visible, infrared, microwave and radar parts of the specturm.

Sensor systems used on aicraft or satellites for detecting radiation are termed *passive* or *active*. *Passive systems* measure radiation energy reflected from objects in the visible and near infrared or energy emitted by objects in the thermal infrared. *Active systems* transmit a beam, part of which is reflected back from objects and recorded by a detector. They operate mainly in the radar region.

Passive remote sensing systems

Passive systems include conventional photography, various scanning systems and multiband spectral photography. Visible light cameras have been used for many years from aircraft and satellites, frequently using overlapping series of black and white or colour photographs. Cameras with special film and filters are also used in the near ultraviolet (0.3–0.4 μm) and in the near infrared (0.7–1.0 μm). These provide very sharp photographs.

Scanning systems, such as the television camera and receiver, rapidly scan a band of the earth's surface as an aircraft carrying the equipment moves along its flight path. Tape-recorded data can then be converted into *imagery*—a picture resembling a photograph.

Multiband spectral photography uses a multiple-lens camera or several cameras; the cameras take photographs of the same area through different filters. The *multispectral scanning system* (MSS) is a sophisticated example carried by the Landsat satellites. It obtains information simultaneously in four bands from 0.5 to 1.1 μm. A scanner feeds the data to detectors after which the data is digitized to a brightness scale and then transmitted to a ground receiving station. The satellite continuously scans a band of 185 km in this way, and on the ground the data are transformed into square photographs representing 185 km × 185 km with 1% overlap between photographs. the system can be refined by presenting three of the four bands in *false colour* (blue, green, red) on the same photograph. False-colour photographs taken in this way show intertidal algae as red, shallow water as blue, and deeper water as dark blue to blue-black.

Scanning radiometers

There are a wide range of scanning systems that rely on *radiometers* in satellites. Radiometers detect radiation reflected or emitted from the sea in different parts of the infrared and microwave electromagnetic radiation spectrum.

Colour scanners are radiometers detecting radiation reflected by the sea in the visible and near infrared bands. The images of ocean colour distribution are used to estimate chlorophyll and phaeopigment concentrations and sediment concentrations. The horizontal resolution is about 1 km. The *Coastal Zone Colour Scanner* (CZCS) is a five-channel scanner aboard the Nimbus 7 satellite which has been regularly used for plotting chlorophyll and phaeopigments in the gulf of Mexico and off South Africa.

Thermal infrared sensors (infra-red scanners) are radiometers that detect radiation from the sea in the 1 to 20 μm range, and do so by using a rotating mirror with a special coating. Infrared radiation is detectable during the day and night and is unaffected by haze or smoke. Warmer areas emit more energy and hence appear darker than cooler areas. Horizontal resolution is 1 to 7 km and temperature accuracies are ± 1 °C. Besides temperature, the intensity of infrared radiation depends on the object's infrared emissivity which is the infrared radiation emitted by the object as a percentage of that emitted by an ideal black body. Most terrestrial emissivities are 85 to 95‰. Thermal infrared sensors can delineate currents in shallow water with different sediment loads seen as light and dark tones.

They can also be used to monitor oil spills, the thermal response being determined by the depth of oil at the sea's surface.

Microwave scanners are radiometers measuring radiation emitted from the sea in the microwave band. The microwave brightness temperatures obtained are used to estimate wind speed, water vapour, rain rate, sea-surface temperature and ice cover. The horizontal resolution is between 15 and 150 km, and wind speed and temperatures can be measured to accuracies of $\pm 2\,\mathrm{ms}^{-1}$ and $\pm 1.5\,^{\circ}\mathrm{C}$ respectively.

Active remote sensing systems

Active systems usually operate in the radar band 0.1–100 cm. *Side-looking airborne radar* (SLAR) transmits impulses either side of an aircraft using a rectangular antenna, and can operate in all weathers except heavy cloud and rain. Surfaces on the earth at right angles to the beam appear light, and those turned away from the beam look dark. SLAR is used for high resolution of land and coastal features.

Altimeters, scatterometers and *synthetic aperture radars* are all micro-wave radars carried on satellites, and are classed as active systems since they transmit a radar beam which is then reflected back to the satellite from the earth's surface.

Altimeters are thin beam radars that accurately measure the vertical distance between the satellite and the earth. They measure the topography and roughness of the sea surface from which the ocean geoid, surface currents and average wave heights can be estimated. Horizontal resolution is 1.5 km, vertical height precision is ± 10 cm, and wave height precision about 50 cm.

Scatterometers also measure the roughness of the sea, but at the same time they record the amplitude of short surface waves that are approximately in equilibrium with the local wind. This enables the velocity of the wind to be calculated. Horizontal resolution is 50 km, and wind speed and direction precision are $1.4\,\mathrm{ms}^{-1}$ and $\pm 18^{\circ}$ respectively.

Polar orbiting and geostationary satellites

Satellites are basically of two sorts: polar orbiting and geostationary. *Geostationary* or *synchronous* satellites are parked in an orbit about 36 000 km above the earth at the equator, and travel at the speed of rotation of the earth's equator (11 000 km h^{-1}). They therefore remain above the same point on the earth. Every 30 minutes, their sensors collect data and imagery of the complete earth disc that they cover. The positions and earth discs of five operational geostationary satellites are shown in Figure 11.4.

Polar orbiting satellites are parked in low orbits between 800 and 1500 km above the earth and circle the earth from 12 to 14 times a day. They obtain imagery and digital data along a band of the earth's surface which can be up to 2500 km.

The *Landsat* satellites are in near-polar orbits at 918 km. Each satellite orbits the earth 14 times a day. Southbound passes are made during the day and northbound ones at night (Figure 11.5, 11.6). The distance between the paths is much greater than the width covered by the *multispectral scanner* (MSS) on board the Landsat. Each day the orbits move slightly westwards, and have repeated themselves exactly on the 19th day. Landsat satellites carry multispectral scanners which are mainly used for land and coastal studies. Since MSSs will not work through cloud and since Landsat only covers the same point once every 19 days, each area on the earth's surface may only be successfully imaged 5 or 6 times a year.

The US National Oceanic and Atmospheric Administration satellites

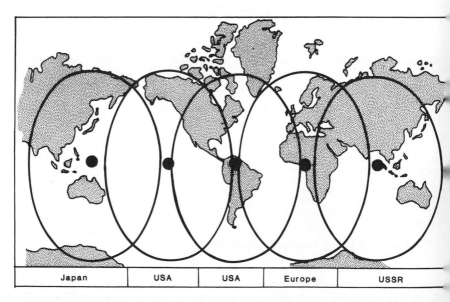

Figure 11.4 Positions and earth discs of geostationary satellites. Japanese satellite: GMS; USA satellites: GOES-W and GOES-E; European satellite: METEOSTAT (after Cracknell, 1982).

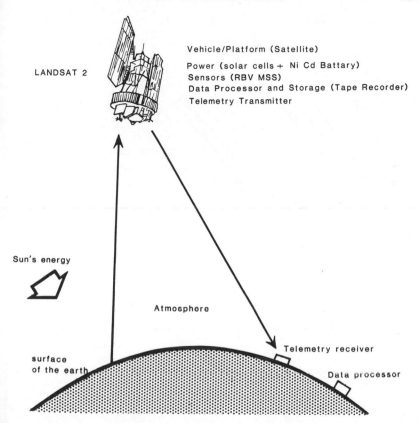

Figure 11.5 Landsat 2 satellite. Diagram of telemetry, data processing and land-based receipt of information (after Lintz and Simonett, 1976).

Tiros-N and *NOAA-6* are also polar orbiting but carry a different scanner, the *Advanced Very High Resolution Radiometer* (AVHRR). AVHRR has a much reduced resolution compared with the MSS on Landsat but covers 1 km rather than about 0.1 km. It has four bands like MSS — but in the thermal infrared — and its pictures are used in TV weather maps. The AVHRR is mainly used for oceanic and meteorological work.

Energy from the oceans

Energy is stored in the oceans in a wide range of forms—winds, waves, currents, tides, temperature and salinity differences, sea bed minerals, and

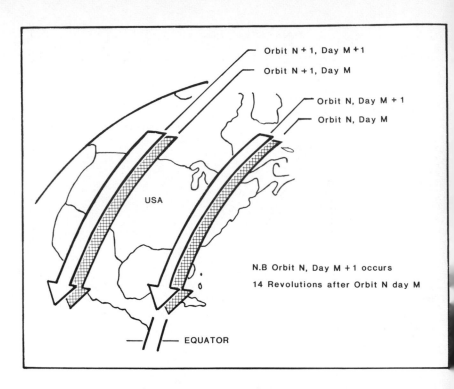

Orbit N + 1, Day M + 1
Orbit N + 1, Day M
Orbit N, Day M + 1
Orbit N, Day M

USA

N.B Orbit N, Day M + 1 occurs
14 Revolutions after Orbit N day M

EQUATOR

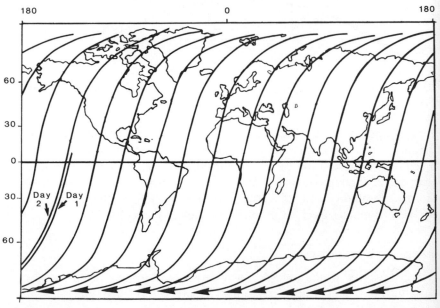

biological productivity. Most of these are renewable and in principle offer almost inexhaustible supplies of energy. However only tides, waves and temperature differences (thermal energy) have been tested commercially.

Tides

Tides have been used in Europe for centuries to drive mill wheels. However, they have only recently been used to produce electricity. There are two site-specific requirements: a tidal range of about 3 m is needed to drive the turbines generating the electricity, and the coastal geography must allow the construction of short dams.

About thirty sites would fulfil these criteria, of which three are in use. The estimated global power of the thirty sites is about 1100 gigawatts ($1\,GW = 10^9\,W$) and so the potential is high. Most of the sites are on the North and South American coasts and in the Middle and Far East (Figure 11.7A).

There are two small experimental schemes — one in China (not indicated in Figure 11.7) and one in Russia (Kislaya Guba on the Barents Sea), and one commercial scheme on the River Rance in France. In addition, three sites are being considered in the Bay of Fundy and one in the Severn Estuary (Table 11.1).

The tidal scheme at La Rance, west of Mont Saint-Michel on the French coast, consists of a concrete barrier across the river and 24 turbines each producing about 540 GWh of electricity per year. Energy is obtained from the ebb and flow of the tide through the turbines. Water can also be pumped into or out of a large basin behind the barrier; the potential energy stored in

Table 11.1 Electricity from tides

Sites	Tidal range	Average power (MW)
A. Operational		
China	not known	3
Kislaya Guba, USSR	2.4 m	20
River Rance, France	8.4 m	240
B. Proposed		
Bay of Fundy, Canada (3 sites)	9.8–10.7 m	1680–19 900
Severn Estuary, UK	9.8 m	7200

Figure 11.6 Upper diagram: LANDSAT ground coverage and flight paths. Lower diagram: Representative LANDSAT ground trace for one day (only southward passes are shown) (after National Aeronautics and Space Administration (NASA) 1972; Lintz and Simonett, 1976).

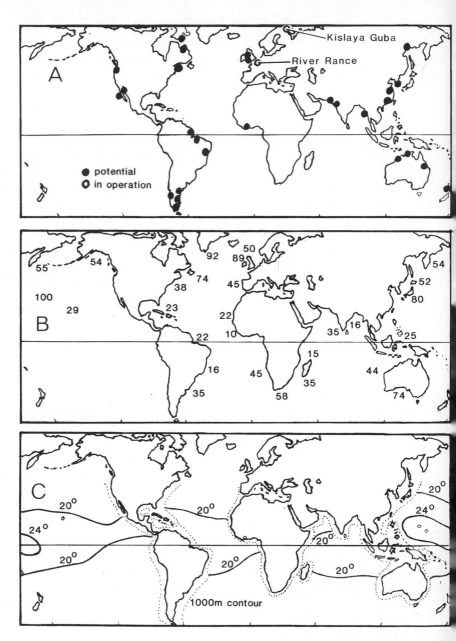

Figure 11.7 Global distribution of ocean energy sources. *A*: tidal energy sites; *B*: wave energy sites (kW. m^{-1} wave front); *C*: Ocean Thermal Energy Conversion devices (OTECs), which must be sited within the 20 °C isotherms and in water deeper than 1000 m (after Couper, 1983).

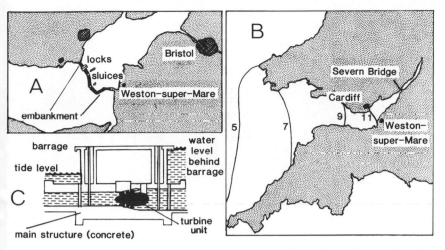

Figure 11.8 Tidal energy. *A*: Proposed inner barrage scheme for the Severn Estuary, Britain. Siting of barrage between Cardiff and Weston-super-Mare. *B*: Tidal range (metres) in the Severn Estuary region showing tidal effect being amplified by the progressive narrowing of the estuary. *C*: Turbine structure used at La Rance tidal power scheme and proposed for the Severn Estuary scheme (after Brown and Skipsey, 1986).

this way can be released when the demand for electricity is high by letting the basin fill or empty by gravity.

An inner barrage scheme was proposed for the Severn Estuary by the Severn Barrage Committee in 1981 (Brown and Skipsey, 1986). This would utilise the funnelling affect of the Severn Estuary, and because of the large water volumes and greater working head, would produce about 30 times the power of La Rance (Figure 11.8, Table 11.1). 160 45-MW turbines would give an installed capacity of 7200 MW or about 1500 MW of continuous power. This is equivalent to about 6% of the present UK electrical energy demand. The scheme would be a simple one with ebb-tide generation only. In spite of this, the environmental impact would be huge. An artificial embankment would stretch across the River Severn from near Cardiff to Weston-super-Mare, and the whole local landscape and shore line would be changed. The scheme would cost about £6000 million and take about 10 years to construct.

Thermal energy

Ocean Thermal Energy Conversion devices (OTECs) use the temperature difference between surface and deep water to produce electricity. They can

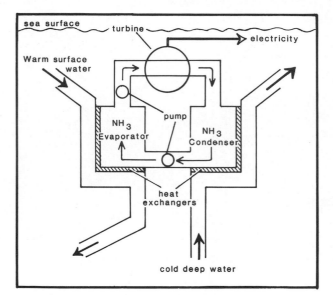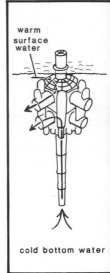

Figure 11.9 Ocean thermal energy conversion devices (OTECs). *Left*: design principle of energy converter at the upper end of an OTEC.
Right: artist's impression of a full-scale OTEC (personal communication, Prof. Berger, University of Hawaii, and after Couper, 1983).

only operate properly in the tropics where the difference in the temperature between the surface and deep waters (1000 to 3000 m) is more than 20 °C. Experimental OTECs are now operating in Japan and the Bahamas, and a MINI-OTEC has also been tested successfully in Hawaii. In principle (Figure 11.9), an OTEC is a 20 m diameter tube extending to at least 500 m depth, at the top of which is a closed container holding liquid ammonia. The liquid ammonia is vaporised to gas by the warm surface water and recondensed to a liquid by the cold water drawn from the deep ocean. The continuous cycle of vaporisation and condensation drives a turbine which produces electricity.

OTECs can be regarded as domestic refrigerators operating in reverse. When a refrigerator is switched on, electricity is used to produce a temperature difference between the inside of the refrigerator which becomes cold, and the outside which becomes slightly warmer. This warmth can usually be felt near the back of a refrigerator. When an OTEC is switched on, the temperature difference existing between the surface and deep waters is used to vaporise and liquefy ammonia which then drives an electric turbine.

Waves

Waves contain a considerable store of energy, especially in winter when they are generally bigger and more frequent. In principle, therefore, their energy could provide electricity at a time of peak demand. Calculations and experimental devices have shown that it is possible to tap this energy, and some remote navigation buoys are already powered using this source. The amount of energy varies from about 10 to $100\,kW\,m^{-1}$ of wave front (Figure 11.7B), and stored either as potential or kinetic energy, rather like a see-saw. Water at the peak has high potential energy; as it moves towards the trough the potential energy is transferred to the kinetic energy of movement. Some of the highest global wave energies occur in the North Pacific ($100\,kW\,m^{-1}$), and also around Rockall and just north-west of the Hebrides, Scotland ($70\text{--}90\,kW\,m^{-1}$) (Figure 11.10). 5 GW, or nearly 20% of the United Kingdom 1985 electricity requirements, could probably be met from this source. Equivalent figures for other countries

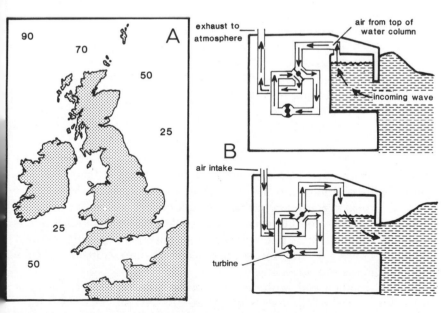

Figure 11.10 A: Wave energies around Britain ($kW\,m^{-1}$ of wave front). B: Conversion of wave energy into electric energy. Operating principle of an oscillating water-column converter designed by the National Engineering Laboratory, UK. Upper diagram: air flow generated by a rising wave. Lower diagram: air-flow generated by a falling wave (after Couper, 1983; Brown and Skipsey, 1986).

(1985) are: Japan 11 GW, 15%; Norway 5 GW, 50%; USA 23 GW, 7% (Brown and Skipsey, 1986).

A number of experimental devices are being tested. Japan and the International Energy Agency have developed a large-scale project called *Kaimei* in which the wave energy is transferred to pneumatic energy, air being forced through a turbine. The air turbine then drives an electrical generator. The Japanese device is 80 m long. Other devices such as the *Lanchester Clam* which also operates by variable air pressure, and the *oscillatory water column* (OWC) systems being developed by the UK National Engineering Laboratory will probably lead to commercial schemes in the 1990s (Figure 11.10).

In December 1986, the Ministry of Oil and Energy in Norway announced that it was studying plans for a 10 MW OWC station based on a design by Kvaerner. The station, which would probably be sited at Bergen, would have 10 concrete OWCs, each 25 m high. Waves enter an opening at the base of the OWC column and set the water inside oscillating. This drives air backwards and forwards through a *Wells turbine* at the top of the tube. These turbines, developed by Professor Wells, Queen's University, Belfast, revolve in the same direction regardless of the direction of the air-flow (Figure 11.10). There have been problems caused by very large waves, but these are being studied by Wells and by a team in Portugal. It is probable that electricity could be supplied by similar schemes in Britain at a cost of about 2.5 pence per unit (1987 prices).

Oil and gas

Oil (petroleum) and gas are found in rocks associated with salt water and sometimes with solid hydrocarbons (coal, bitumens), and there is now a global industry drilling for them on the world's continental shelves.

Oil is formed in marine, estuarine or deltaic environments. Its formation is not fully understood but it is associated with the microbiological and chemical decay of plants and animals under anaerobic conditions which takes place during the transformation of sediments into sedimentary rocks beneath the sea bed (diagenesis).

After formation, oil with gas in solution moves readily through permeable rocks and sediments, progressively separating out as a layer above the salt water in the interstices of the rock. The lower boiling-point hydrocarbons may separate as a natural gas layer above the oil, and higher boiling-point hydrocarbons may solidify as waxes and asphalts. The oil migrates predominantly upwards until it reaches the surface of the rock or

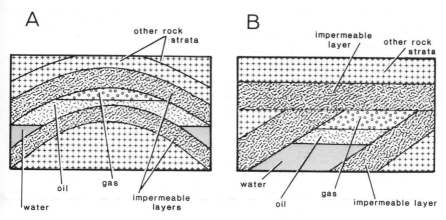

Figure 11.11 Oil accumulation in rocks. *A*: anticlinal fold. *B*: unconformity (after Whitten and Brooks, 1978).

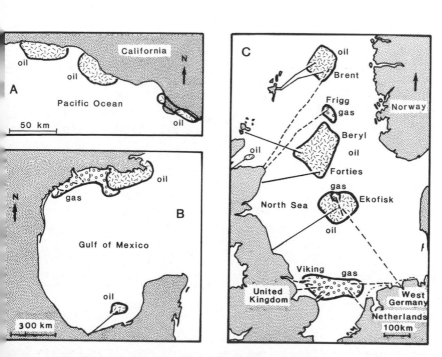

Figure 11.12 Oil and gas deposits: *A*, off California; *B*, in the Gulf of Mexico; *C*, in the North Sea. ——oil, and - - - - gas pipe lines (after Couper, 1983).

sediment or is trapped beneath impervious rock layers. These *rock traps* can be either structural, formed by faulting or folds, or stratigraphic, formed by unconformities or changes of *facies* (Figure 11.11).

The oil and gas trapped in the permeable rock is called an *oil pool* although it is still in the rock's interstices. Once the oil pool is drilled the pressure of the salt water in the rocks beneath the pool is normally enough to drive the gas and oil upwards. Natural gas can occur without oil in these rock layers, but not vice versa.

The first marine exploration for oil was in 1896, when drilling was undertaken from wooden piers to exploit a submarine extension of the Summerland field in Southern California. Since then, marine exploration and exploitation has become worldwide. The world's major offshore oil and gas producing countries are Abu Dhabi, the USSR, Saudi Arabia, the United Kingdom, the United States and Venezuela (off the Caribbean Sea). The major areas where oil is commercially exploited include the Gulf of Mexico, Lake Maracaibo, Venezuela, the North Sea, and the Persian Gulf (Figure 11.12). Recovery of oil from the sea bed is done firstly by exploration rigs such as semi-submersibles and drilling ships, and then by

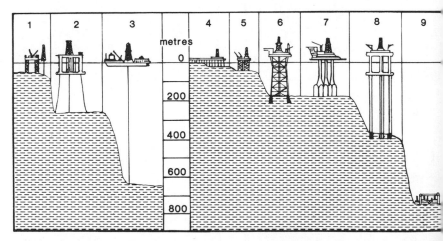

Figure 11.13 Drilling techniques for oil and gas (after Couper, 1983).
Left: offshore exploration drilling. 1, Jackup rigs; 2, semisubmersible platforms; 3, drill ships.
Right: Continental shelf (4–7) and deep-sea (8, 9) commercial drilling, 4, extension of onshore drilling; 5, steel-framed structures directly piled into the sea bed (< 50 m); 6, steel and 7, concrete platforms built ashore and then towed to site; 8, a hyperbuoyant tension leg platform and 9, an undersea manifold used beyond the edge of the continental shelf (9 needs a separate drilling production well).

oil production rigs, the design of which depends largely on water depth (Figure 11.13).

The exploitation of new areas now follows a fairly predictable course. Site exploration by geological surveys and drilling is followed by development of land service bases and engineering yards to build production platforms. Finally, before full-scale production can start, pipelines on land and on the sea bed, and coastal, oil and gas terminals are needed. The whole sequence requires a complex and efficient industrial structure which is only available in the large national and multinational oil companies. Once oil and gas are on line, the offshore equipment and rigs need continuous service and facilities which include helicopters, surface supply vessels, accommodation ships, trenching and pipe-laying ships and standby fire-fighting vessels. There is also a great deal of subsurface activity—SCUBA divers, oxyacetylene welding, diving bells, small submarines, remote-control TV, storage tanks and so on.

The North Sea was the most important area for the development of offshore and deeper water engineering techniques in the 1960s and 1970s. It now contains a network of commercial operating oil and gas rigs centred on four areas: Northern (Brent, Murchison), Northern and Central (Frigg, Beryl); Central (Ekofisk, Tor) and Southern (Leman Bank, Viking) (Figure 11.12).

The production rig itself is usually a steel structure piled into the sea bed, with a steel deck above sea level for the production equipment and accommodation. The rigs are often grouped together so that costs can be reduced by sharing transportation and pipeline facilities. Oil is taken from the rigs to land by sea-bed oil pipes or stored in large concrete storage tanks which can hold up to 1 million barrels where water depth is greater than 200 m. Gas is always carried by pipelines.

As oil exploitation moves off the continental shelf into the deep sea, the whole operation is likely to be conducted from floating submarine platforms (*subsea completions*) anchored near the sea bed.

Sea-bed minerals

Sea-bed minerals on the continental shelf such as sand and gravel have been mined for many years, but there is now an interest in deep water deposits such as tin in submerged *placer deposits*, mineralised muds in the Red Sea, and manganese nodules and metal-rich sediments in many deep sea basins (Figure 11.14).

At present (1987) the most important sea-bed minerals commercially are

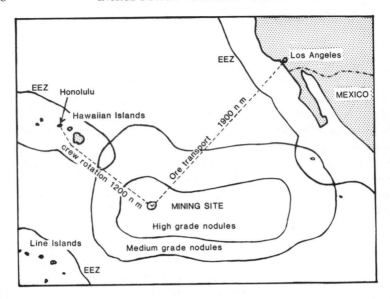

Figure 11.14 North-East Pacific manganese nodule zone and hypothetical mining site in a high-grade nodule area between Hawaii and Los Angeles (after Couper, 1983).

sand, gravel and tin, and there is a well established range of dredging techniques. *Sand* and *gravel* are used mainly in the construction industries. On beaches they are formed from cliff erosion or derived from rivers. On the continental shelf they are derived partly from rivers but mainly from relict sediments from the Quaternary era or earlier. In deeper waters *calcareous sediments* are made up of fragmented shells and precipitation from sea water. Two of the most important commercially mined areas for calcareous sediments are shell deposits off Iceland and oolites (aragonite) on the Bahamas Banks.

A number of factors determine the commercial viability—grain size, composition, production and transport costs and depth of deposits. Dredging activities offshore may also conflict with gas and oil interests. There are important sand and gravel deposits on the North American coasts, and also in the North Sea, which is very rich and has been divided into concession areas. Estimates for the North American continental margins are about 1400×10^9 tonnes.

Placer deposits are sediments containing minerals such as casserite (tin), magnetite (iron), and rutite (titanium) eroded from ore bodies by mechanical weathering, mainly over the last 65 million years. A number of beaches have placer deposits—titanium sands in Florida, Sri Lanka and Brazil, tin in Malaysia and Indonesia, and magnetite in Japan. Offshore, one of the main sites is Indonesia, where exploration is occuring to 50 m at present.

The importance of *metalliferous sediments* has only been appreciated since the 1960s. Manganese and iron in sediments have been recorded by the Deep Sea Drilling Project along the East Pacific Ridge and many island arcs. However, these are sometimes buried deep beneath other sediments and therefore are not easily accessible. *Polymetallic sulphide deposits* have also been recorded in hydrothermal vent areas on the North Pacific Ridge where they form cylindrical structures 3 to 4 m high and about 3 m in diameter.

The Red Sea contains a number of deep basins below 1000 m in which there are *metalliferous muds*. The muds consist of multicoloured layers of very fine sediment containing up to 1% Cu, 6% Zn and 100 ppm Ag. They are between 2 and 25 m thick and are covered by about 200 m of high-salinity water at temperatures up to 60 °C. The temperature and the multilayering suggest continuing hydrothermal activity in the area. The 2000 m Atlantis II Deep in the centre of the Red Sea north-west of Jeddah covers about 60 km². It is in Sudan's and Saudi Arabia's Exclusive Economic Zones, and a joint commission administers its exploration and exploitation. The Atlantis II Deep is likely to become commercially viable by the 1990s.

Phosphatic deposits, also called *phosphorites* after the mineral *phosphorite* (calcium phosphate), are sediments and rocks of marine origin containing significant amounts of phosphate (6–33% P_2O_5 equivalent). Phosphates are a major requirement of the fertiliser industry, and phosphatic deposits are commercially mined on land by Morocco and the USA. At present, however, there is no commercial mining of sea-floor deposits although attempts were made off California in the 1960s. Phosphorites are found as muds, sand nodules (pebble to boulder size) and pavements, and are known from the continental margins off North and South America, Africa, Spain, and Australasia, as well as from a number of sea mounts in the Pacific and Atlantic. They are usually found at depths of less than 500 m between 40° N and 40° S. The mechanism of their formation is not fully understood. Phosphorites are strongly associated with upwelling at continental margins; on the other hand, diatom ooze, a widespread deep sea sediment, can contain up to 4% P_2O_5. The accumulated excreta of sea birds, *guano*, found on some oceanic islands and shore cliffs, is another source of phosphate, and being on land is easily mined.

Manganese nodules, with their high content of valuable metals (Mn, Ni, Cu, Co), are an obvious target for marine mining. The problem is depth, since most are found below 4000 m. The only area at present feasible for mining is between the Clarion and Clipperton fracture zones, extending west from the Californian coast to the Hawaiian Islands and the Line Islands (Figure 11.14). Here the nodules may cover more than 50% of the sea floor. The area is thought to contain about 17×10^6 tonnes of copper, 2.5×10^6 tonnes of cobalt, and 300×10^6 tonnes of manganese. These are very large quantities, and would extend the world's resources significantly. However the problems of mining even this area are enormous because of water depths, ore transport, crew, and spare part replacements. On site, the mining of manganese nodules involves two large ships, and the riser pipe from the sea bed needs flotation chambers with auxiliary pumps along its length. In addition, a riser pipe lifting 10 000–15 000 tonnes of nodules per day needs about 4 MW of electric power. The sea-floor harvesting vehicle is either self-propelled or towed, and collects the nodules by a 'vacuum cleaner' device or heaps them into lines before they are brought to the surface by a batch lift system.

The size of a seabed concession for a profitable manganese nodule mining operation over 20 years is about 50 000 km². Estimates based on samples from about 2000 sites globally suggest that only 25 to 100 are of this size. Hence there is very high commercial competition.

Sulphur is important in many industrial processes including fertiliser production. Natural gas is the main source, followed by sulphur in the cap rock of deep salt domes. Salt domes form in rocks under high pressure when sodium chloride deforms plastically and behaves like an intrusive magma (fluid rock). They sometimes deform and even pierce the overlying sediments and may contain oil.

Minerals from sea water

There are a large number of elements in sea water which could be extracted, but apart from sea salt the only major commercial extraction is that of magnesium and bromine. Magnesium is obtained by preparing $MgCl_2$ from sea water, followed by electrolysis to give the pure metal. Bromine is obtained by treating sea water with sulphuric acid and chlorine which releases bromine. Magnesium, being very light, is used in a wide range of industrial application, and bromine is used in anti-knock compounds in petrol.

In the future the extraction of uranium for use by the nuclear industries and of deuterium and hydrogen for nuclear fission may be a viable proposition.

Fresh water

Fresh water obtained from the sea is used for drinking and for agriculture in some tropical and subtropical countries; it is also used in industry. It is extracted from sea water where natural fresh water is limited. The most important method is desalination, but distillation and freezing are also used. *Reverse osmosis* is a new method which is now used on ships to provide a constant supply. For example, the British research vessel *Challenger* produces about 14 000 l of pure water per day by reverse osmosis, which is well in excess of that needed for drinking, washing cooking, and for scientific and engineering purposes.

Salt

Salt has been obtained from sea water for about 4000 years by solar evaporation from salt pans. This is most effective in hot countries with a low humidity, low rainfall and high evaporation, and in the tropics and subtropics it is fairly common. Modern methods of extracting sea salt include a series of ponds that remove mud and calcium sulphate before concentrating the sodium chloride.

Pollution

Pollution in the sea is a global problem ranging in scale from major oil spills to the indiscriminate disposal of plastic bags by holiday makers on beaches. It can be continuous (sewage) or episodic (the Amoco Cadiz wreck, North Brittany 1978), and deliberate (waste disposal) or accidental (oil spills). Some pollutants are naturally-occurring substances (domestic sewage) while others are man-made (DDT).

   Dredging of estuarine and inshore ship channels and resultant dumping causes increased turbidity and some destruction of benthic ecosystems, and this can also be considered as a form of pollution.

Advantages, routes of entry, types, persistence

There are advantages to controlled waste disposal at sea. Natural wastes are readily degraded, toxic materials are rapidly diluted to low levels, and solids are permanently lost into sediments. In addition, waste disposal at sea is usually cheaper and safer than on land. Sewage farms often utilise expensive land near towns, for example, and toxic wastes are a health hazard on local rubbish tips. The disadvantages of disposal at sea are obvious: some pollutants are highly toxic even at low levels (heavy metals), insecticides are concentrated up food chains and can kill marine birds and mammals, and some radioactive materials are virtually permanent (Figure 11.15).

Pollutants enter the sea by many routes. Land sources are common: sewage from coastal towns and industries often drains directly into estuaries or the sea, land reclamation usually includes waste tipping on shorelines, and regular agricultural use of fertilisers (nitrate, phosphate) and pesticides (DDT, PCBs) leads to polluted run-off flowing into rivers and on to shores.

Heavy metals (lead, mercury, copper, arsenic, cadmium, selenium) are broken down by chemical changes. Lead and mercury change to more poisonous methyl compounds, and many heavy metals are absorbed and concentrated by marine animals. Occasionally man is affected. Methyl

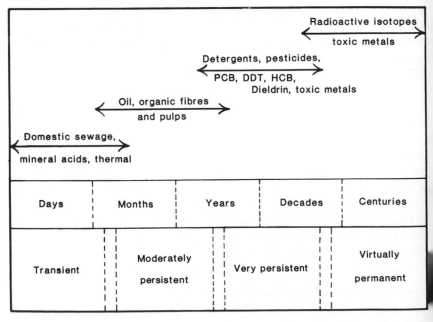

Figure 11.15 Persistence of pollutants in the sea (after FAO, 1977; Geyer, 1980; Clarke, 1986).

mercury infected fish in Minamata Bay, Japan, in the 1950s, and the members of the local fishing community reported odd symptoms such as loss of coordination, speech impairment and numbness of the limbs.

Sewage is either discharged untreated or partly treated at or just below low water, or is dumped near coasts. It contains phosphates, nitrates and ammonia which upset local marine productivity. However, the main dangers, particularly from domestic sewage, are pathogenic bacteria and viruses causing diseases such as *Salmonella* food poisoning, typhoid, and hepatitis. Man is infected either directly while bathing or by eating contaminated shellfish.

At sea, controlled dumping of sewage and radioactive materials is legal. Most ships still tip waste over the side, tank-washing by oil tankers is common, and marine dumping of various materials occurs. In addition, accidental collisions, wrecks and pipeline damage are regular events. Radioactive pollution at sea has come from global fall-out from nuclear testing (banned since 1963), from dumping packaged waste in deep water, and from leakage from coastal nuclear plants (Sellafield, Britain).

Oil is transported in very large volumes by tankers at sea. Intentional discharge of oil or oily water is now illegal, but accidental release often happens after collisions or groundings. The break-up of the *Amoco Cadiz* oil tanker off the coast of Brittany in 1978 is an example. Hundreds of thousands of tons of oil may be released, and if near land they cause widespread inshore pollution and death of intertidal animals and plants, especially birds. At least 100 000 birds die each year on British coasts from oil. The effect of large spills of oil is sometimes unexpected, however. On 29th March 1957 the oil tanker *Tampico Maru* was wrecked on the Pacific coast of Mexico, 180 km south of the Mexico/US border. The tanker blocked about three-quarters of the entrance of a small cove, which caused a marked reduction of wave action within the cove. Oil from the tanker sporadically leaked out until the wreck broke up some 6 to 8 months later, and caused a heavy mortality of many benthic animals including the sea urchin *Strongylocentrotus franciscanus* and the abalone *Haliotis cra- cherodii*, both of which graze seaweeds. At the same time, the seaweed *Macrocystis pyrifera* became very much more abundant in the cove, its abundance only returning to pre-wreck levels some 6 years later Figure 11.16). This dramatic increase in *Macrocystis* is thought to have been caused partly by the reduction in wave action within the cove, but mainly by the mass mortality of the sea urchins and abalones that normally graze on the seaweed fronds (North, Neushul and Clendenning, 1964).

The treatment of oil spills is a subject which is being actively investigated

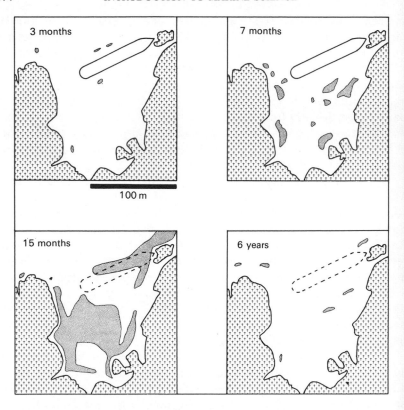

Figure 11.16 Progressive growth and subsequent disappearence of the seaweed *Macrocysti*
in a sheltered bay on the Pacific coast of Mexico, after the wreck of the oil tanker *Tampic*
Maru on 29th March, 1957. The extent of *Macrocystis* growth is represented by deeply stipple
areas in the bay, and is shown at its greatest after 1 year 3 months (bottom left) (modified fror
North, Neushul, and Clendenning, 1964).

at the present time. The oil usually floats at the water surface, but is ofte
difficult to remove. Detergents disperse it, but probably kill more animal
than does the oil, particularly on the sea shore. High-density powder such a
chalk has been used to sink oil, but this treatment affects the bottom faun
and fouls fishing gear, and the oil may also return to the surface. Perhap
one of the most successful techniques recently developed has been the use c
large flexible buoyant booms to contain local spills. These are laid in a circl
around the patch and the surface layer of oil then pumped on board
recovery ship.

Pollutants can also enter the sea from the atmosphere as dust and b

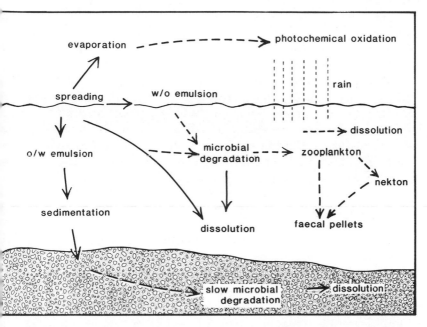

Figure 11.17 Progressive degradation of oil after an oil spill at sea (after FAO, 1977).

solution of volatile materials in rain (CO, SO$_2$), thus forming acid rain. These can persist in the sea for days to decades depending on the material, and undergo a series of changes during the process of breakdown Figure 11.17). For example, after an oil spill 12-carbon and then 22-carbon compounds in oil immediately evaporate at the air/sea interface and photo-oxidise or auto-oxidise in air, eventually returning to the sea. At the same time water-in-oil (w/o) and oil-in-water (o/w) emulsions form which are then degraded by microorganisms. Oil breakdown products also enter food chains via zooplankton, and become incorporated into sediments.

The global pollution problem

Pollution occurs to some extent in all the major oceans; however, in some, such as the Mediterranean, it is becoming a major health hazard. The Mediterranean has a limited exchange of its water—once every 70 years—and is bordered by 18 states containing about 100 million people.

The north Mediterranean coast, especially of Spain, France and Italy, is heavily polluted by oil and sewage, while the southern coast, particularly

Algeria and Libya, is less so. In 1975 the states bordering the Mediterranean adopted a *Mediterranean Action Plan* coordinated under the United Nations Environmental Programme (Regional Seas). This plan consists of an inventory of all pollution sources and proposals for action.

Coastal and inshore waters that are near large centres of human population on coasts are polluted by a wide range of materials. The North Sea, the Irish Sea and the Baltic are the most polluted marine areas in the world. Sewage, industrial waste and radioactive pollution are at dangerous levels particularly near English, French, Dutch and German coasts and in the English Channel. In Britain, the high level of radioactive material in the Irish Sea near the nuclear power station at Sellafield has become a scandal.

There are also major areas of pollution in the Middle and Far East and in Australasia. Heavy oil pollution occurs in the Persian Gulf (Saudi Arabia, Iraq, Iran), heavy metal pollution in Bombay Harbour and massive pollution at the mouth of the Ganges (India). Epidemics of typhoid and hepatitis from shellfish are regularly reported on Indonesian and Vietnamese coasts, and industrial waste has badly affected Chinese and Japanese local waters. Off Sydney, Australia, there is a health risk to bathers from inadequate sewage treatment, and in New Zealand waste from animal product factories (milk, meat) cause local problems.

The coasts of North and South America are heavily polluted locally, in particular those of the United States and Canada. Radioactive waste from uranium mining has affected salmon in British Columbia; dumping of sewage-treated sludge off New York and New Jersey has caused local fish mortalities and oxygen deficiencies over $12\,000\,km^{-2}$; heavy oil pollution is present in Lake Maracaibo (Venezuela); sewage pollution off Rio de Janeiro and Santos (Brazil) is heavy; and highly toxic industrial waste in the River Plate has caused fish mortalities and prohibition of swimming around Buenos Aires.

In contrast to this sad story of man's abuse of nearshore and continental shelf waters, the centres of the oceans in the northern hemisphere and most waters in the southern hemisphere are relatively less polluted. It is to be hoped that they remain so.

BIBLIOGRAPHY

References

Alexander, M. (1971) *Microbial Ecology*. Wiley, London.

Alexander, W.B., Southgate, B.A., and Bassindale, R. (1935) Survey of the River Tees. Part II– The Estuary–Chemical and Biological. *D.S.I.R. Water Pollution Research Technical Paper No. 5*.

Allen, J.R.L. (1968) *Current Ripples*. North-Holland, Amsterdam.

Allen, J.R.L. (1970) *Physical Processes of Sedimentation*. Allen and Unwin, London.

Allen, J.R.L. (1985) *Principles of Physical Sedimentology*. Allen and Unwin, London.

Aller, R.C. (1983) *The importance of the diffusive permeability of animal burrow linings in determining marine sediment chemistry*. *J. Marine Res.* **41**, 299–322.

Anderson, H.T. (ed.) (1969) *The Biology of Marine Mammals*. Academic Press, New York.

Anderson, J.G. and Meadows, P.S. (1978) Microenvironments in marine sediments, *Proc. Roy. Soc. Edinb.* **B76**, 1–16.

Anon. (1952) *Marine Fouling and its Prevention*. Prepared for the Bureau of ships, Navy Department, by Woods Hole Oceanographic Institution. United States Naval Institute: Annapolis, Maryland.

Anon. (1971) *Radioactivity in the Marine Environment*. National Academy of Sciences, Washington.

Anon. (1972) *Atlas of the Living Resources of the Seas*. FAO, Rome.

Anon. (1975) *Numerical Models of Ocean Circulation*. National Academy of Sciences, Washington.

Baas Becking, L.G.M., Kaplan, I.R., and Moore, D. (1960) Limits of the natural environment in terms of pH and oxidation-reduction potentials. *J. Geol.* **68**, 243–284.

Bainbridge, R. (1957) The size, shape and density of marine phytoplankton concentrations. *Biol. Revs.* **32**, 91–115.

Baker, A. deC., Clarke, M.R., and Harris, M.J. (1973) The N.I.O. combination net (RMT 1 + 8) and further developments of rectangular midwater trawls. *J. Marine Biol. Assn. of the United Kingdom* **53**, 167–184.

Baker, J.T. and Murphy, V. (eds.) (1974) *Handbook of Marine Science—Compounds from Marine Organisms*. 2 vols. CRC Press, Cleveland.

Baker, J.M. (ed.) (1976) *Marine Ecology and Oil Pollution*. Applied Science, Barking, Essex.

Bambach, R.K. (1977) Species richness in marine benthic habitats throughout the Phanerozoic. *Paleobiology* **3**, 152–157.

Banse, K. and Mosher, S. (1980) Adult body mass and annual production/biomass relationships of field populations. *Ecol. Monogr.* **50**, 355–379.

Bardach, J.E., Ryther, J.H. and McLarney, W.O. (1972) *Aquaculture*. Wiley–Interscience, New York.

Barnes, H. (1959) *Oceanography and Marine Biology*. Allen and Unwin, London.

Barnes, H. (ed.) (1966) *Some Contemporary Studies in Marine Science*. Allen and Unwin, London.

Barnes, R.S.K. (1984) *Estuarine Biology*. 2nd edn. Edward Arnold, London.

Barnes, R.S.K. and Green, J. (eds.) (1972) The estuarine environment. In *1st Symp. of the Estuarine and Brackish-water Biological Association*. Applied Science, London.

Barnes, R.S.K. and Hughes, R.N. (1982) *An Introduction to Marine Ecology*. Blackwell, Oxford.

Barnes, R.S.K. and Mann, K.H. (eds.) (1980) *Fundamentals of Aquatic Ecosystems*. Blackwell, Oxford.

Barnett, P.R.O. (1968) Distribution and ecology of harpacticoid copepods of an intertidal mud flat. *Int. Rev. der gesamten Hydrobiologie und Hydrographie* **53**, 177–209.

Barrett, E.C. and Curtis, L.F. (1976) *Introduction to Environmental Remote Sensing*. Chapman and Hall, London.

Bascon, W. (1960) Beaches. *Scientific American* **203**, 80–94.

Battaglia, B. and Beardmore, J.L. (eds.) (1978) *Marine Organisms: Genetics, Ecology and Evolution*. Plenum, New York.

Bayne, B.L. (ed.) (1976) *Marine Mussels: Their Ecology and Physiology*. Cambridge University Press, Cambridge.

Belderson, R.H., Kenyon, N.H., Stride, A.H. and Stubbs, A.R. (1972) *Sonographs of the Sea Floor*. Elsevier, Amsterdam.

Beverton, R.J.H. and Holt, S.J. (1957) On the Dynamics of Exploited Fish Populations. *Ministry of Agriculture, Fisheries and Food: Fishery Investigations*, Series II, Vol. XIX., HMSO, London.

Bird, C.F.E. (1984) *Coasts. An Introduction to Coastal Geomorphology*. Blackwell, Oxford.

Boney, A.D. (1966) *A Biology of Marine Algae*. Hutchinson, London.

Boney, A.D. (1975) *Phytoplankton*. Edward Arnold, London.

Borgese, E.M. and Ginsburg, N. (eds.) (1982) *Ocean Yearbook 3*. University of Chicago Press, Chicago and London.

Bougis, P. (1976) *Marine Plankton Ecology*. North-Holland, Amsterdam.

Bowman, M.J. and Easias, W.E. (eds.) (1978) *Oceanic Fronts in Coastal Processes*. Springer Verlag, Berlin.

Brafield, A.E. (1978) *Life in Sandy Shores*. Edward Arnold, London.

Brauer, R.W. (ed.) (1972) *Barobiology and the Experimental Biology of the Deep Sea*. University of North Carolina Press, Chapel Hill.

Brewer, P.G. (ed.) (1983) *Oceanography. The Present and Future*. Springer Verlag, New York, Heidelberg.

Brinkhuis, B.H., Tempel, N.R. and Jones, R.F. (1976) Photosynthesis and respiration of exposed salt-marsh fucoids. *Marine Biol.* **34**, 349–359.

Brock, T.D. (1966) *Principles of Microbial Ecology*. Prentice-Hall, New Jersey.

Broecker, W.S. (1974) *Chemical Oceanography*. Harcourt Brace Jovanovich, New York.

Brown, G.C. and Skipsey, E. (eds.) (1986) *Energy Resources. Geology, Supply and Demand*. Open University Press, Milton Keynes/Philadelphia.

Brunn, P. (1966) *Tidal Inlets and Littoral Drift*. Universitetsforlaget, Norway.

Buzzati-Traverso, A.A. (ed.) (1960) *Perspectives in Marine Biology*. University of California Press, Berkeley and Los Angeles.

Burk, C.A. and Drake, C.L. (eds.) (1974) *The Geology of Continental Margins*. Springer Verlag, New York.

Cameron, A.M. et al. (eds.) (1974) *Proceedings of the 2nd International Symposium on Coral Reefs*. (2 vols). The Great Barrier Reef Committee, Brisbane.

Cameron, W.M. and Pritchard, D.W. (1963) Estuaries. In *The Sea*, vol. 2 (M.N. Hill, ed.). Wiley Interscience, New York.

Campbell, J.I. and Meadows, P.S. (1974) Gregarious behaviour in a benthic marine amphipod (*Corophium volutator*). *Experientia* **30**, 1396–1397.

Capuzzu, J. (ed.) (1987) *Biological Processes and Waste in the Ocean. Vol. 1. Oceanic Processes in Marine Pollution*. Krieger, Malabar, Florida.

Carefoot, T. (1977) *Pacific Seashores. A Guide to Intertidal Ecology*. Douglas, Vancouver.

Carrington, R. (1960) *A Biography of the Sea*. Chatto and Windus, London.
Carthy, J.D. and Arthur, D.R. (eds.) (1968) *The Biological Effects of Oil Pollution on Littoral Communities*. Field Studies Council, London.
Chapman, A.R.O. (1986) Population and community ecology of seaweeds. *Adv. Marine Biol.* **23**, 1–161.
Chapman, V.J. (1970) *Seaweeds and their Uses* (2nd edn.) Methuen, London.
Chapman, V.J. (1975) *Mangrove Vegetation*. Lehre J. Cramer.
Chapman, V.J. (1977) *West Coast Ecosystems*. Elsevier, Amsterdam.
Chapman, V.J. (ed.) (1977) *Ecosystems of the World 1: Wet Coastal Ecosystems (Salt Marshes, Mangrove Swamps)*. Elsevier, Amsterdam.
Chapman, V.J. and Chapman, D.J. (1973) *The Algae*. (2nd edn.) Macmillan, London.
Chapski, K.K. and Sokolov, V.E. (1973) (transl. H. Mills) *Morphology and Ecology of Marine Mammals*. Wiley, London.
Clarke, R.B. (1986) *Marine Pollution*. Oxford University Press, Oxford.
Cole, H.A. (ed.) (1979) *The Assessment of Sublethal Effects of Pollutants in the Sea*. A Royal Society Discussion. The Royal Society, London.
Colwell, R.R. and Morita, R.Y. (eds.) (1974) *Effect of the Ocean Environment on Microbial Activities*. University Park Press, Baltimore.
Colwell, R.R. *et al.* (10 authors) (1975) *Marine and Estuarine Microbiology Laboratory Manual*. University Park Press, Baltimore.
Connell, J.H. (1961) The influence of interspecific competition and other factors on the distribution of the barnacle *Chthamalus stellatus*. *Ecology* **42**, 710–723.
Costerton, J.W. and Colwell, R.R. (eds.) (1979) *Natural Aquatic Bacteria: Enumeration, Activity and Ecology*. American Society for Testing and Materials.
Costlow, J.D. and Tipper, R.C. (eds.) (1984) *Marine Biodeterioration: An Interdisciplinary Study*. Proceedings of the Symposium on Marine Biodeterioration, Uniformed Services University of Health Sciences. E. & F.N. Spon, London.
Coull, B.C. (ed.) (1977) *Ecology of Marine Benthos*. University of South Carolina Press, Columbia.
Couper, A. (ed.) (1983) *The Times World Atlas of the Oceans*. Times Books, London.
Cracknell, A.P. (ed.) (1981) *Remote Sensing in Meteorology, Oceanography, and Hydrography*. Ellis Horwood, Chichester.
Cracknell, A.P. (ed.) (1982) *Remote Sensing Applications in Marine Science and Technology*. Reidel, Dordrecht.
Craig, R.E. (1972) *Marine Physics*. Academic Press, London.
Crisp, D.J. (ed.) (1964) *Grazing in Terrestrial and Marine Environments*. Blackwell Scientific, Oxford.
Crisp, D.J. and Meadows, P.S. (1962) The chemical basis of gregariousness in cirripedes. *Proc. Roy. Soc. London* **B156**, 500–520.
Crisp, D.J. and Meadows, P.S. (1963) Adsorbed layers: the stimulus to settlement in barnacles. *Proc. Roy. Soc. London* **B158**, 364–387.
Cronan, D.S. (1980) *Underwater Minerals*. Academic Press, London.
Cronin, L.E. (ed.) (1975) *Estuarine Research* (2 vols.) Academic Press, New York.
Cross, M.G. (1978) *Oceanography: A View of the Earth*. (2nd edn.) Prentice-Hall, New Jersey.
Cushing, D.H. (1968) *Fisheries Biology*. University of Wisconsin Press, Modison.
Cushing, D.H. (1973) *The Detection of Fish*. Pergamon Press, London.
Cushing, D.H. (1975) *The Productivity of the Sea*. Oxford Biology Readers No. 78, Oxford University Press, London.
Cushing, D.H. (1977) *Science and the Fisheries*. (Studies in Biology no. 85). Edward Arnold, London.
Cushing, D.H. and Walsh, J.J. (eds.) (1976) *The Ecology of the Seas*. Blackwell, Oxford.
Davis, R.A. (1978) *Principles of Oceanography*. (2nd edn.) Addison-Wesley, Massachusetts.
Day, J.H. (1969) *A Guide to Marine Life on South African Shores*. Balkema, Cape Town.

Dayton, P.K. (1973) Dispersion, dispersal, and persistence of the annual intertidal alga, *Postelsia palmaeformis* Ruprecht. *Ecology* **54**, 433–438.

Dayton, P.K. (1985) Ecology of kelp communities. *Ann. Rev. Ecol. and Systematics* **16**, 215–245.

Deepak. A. (ed.) (1980) *Remote Sensing of Atmospheres and Oceans.* Academic Press, New York.

Defant, A. (1961) *Physical Oceanography* (2 vols.) Pergamon Press, Oxford.

den Hartog, C. (1970) *The Seagrasses of the World.* North Holland, London.

Dennes, B. (ed.) (1984) *Seabed Mechanics.* Proceedings of a Symposium, sponsored jointly by the International Union of Theoretical and Applied Mechanics (IUTAM) and the International Union of Geology and Geophysics (IUGG). Graham and Trotman, London.

Denny, M.W. (1987) Life in the maelstrom: the biomechanics of wave-swept rocky shores. *Trends in Ecology and Evolution* **2**, 61–66.

Dixon, P.S. (1973) *Biology of the Rhodophyta.* Oliver and Boyd, Edinburgh.

Donovan, D.T. (ed.) (1968) *Geology of Shelf Seas.* Oliver and Boyd, Edinburgh.

Drake, C.L., Imbrie, J., Knauss, J.A. and Turekian, K.K. (1978) *Oceanography.* Holt Rinehart and Winston, New York.

Dunbar, M.J. (ed.) (1979) *Marine Production Mechanisms.* Cambridge University Press: Cambridge.

Duxbury, A.C. (1971) *The Earth and its Oceans.* Addison Wesley, Reading, Mass.

Dyer, K.R. (1973) *Estuaries: a Physical Introduction.* Wiley, Chichester.

Dyer, K.R. (ed.) (1979) *Estuarine Hydrography and Sedimentation.* Cambridge University Press, Cambridge.

Edmunds, P.J. and Spencer Davies, P. (1986) An energy budget for *Porites porites* (Scleractinia). *Marine Biol.* **92**, 339–347.

Ekman, S. (1953) *Zoogeography of the Sea.* Sidgwick and Jackson, London.

Elliot, F.E. (1960) General Electric Advanced Electronics Center at Cornell University Rept No. R 60EL C45.

Eltringham, S.K. (1971) *Life in Mud and Sand.* English Universities Press, London.

Emery, K.O. (1960) *The Sea off Southern California.* Wiley, Chichester.

Emery, K.O. (1968) Relict Sediments on the Continental Shelves of the World. *Bull. Amer. Assn. Petrol. Geol.* **52**, 445–464.

Emery, K.O. (1969) The continental shelves. *Scientific American* **221**, 106–122.

Emery, K.O. and Uchupi, E. (1972) Western North Atlantic Ocean: Topography, Rocks, Structure, Water, Life and Sediments. *Association of Petroleum Geologists Memoir* 17.

Estes, J.A. and Palmisano, J.F. (1974) Sea Otters: their role in structuring nearshore communities. *Science* **285**, 1058–1060.

FAO (1977) *Impact of Oil on the Marine Environment.* Rep. Stud. FAO, Food and Agriculture Organisation, Rome.

Fenchel, T.M. (1978) The ecology of micro- and meiobenthos. *Ann. Rev. Ecol. and Systematics* **9**, 99–121.

Fenchel, T.M. (1987) *Ecology of Protozoa.* Science Tech/Springer Verlag, Madison/Berlin.

Firth, F.E. (ed.) (1969) *The Encyclopaedia of Marine Resources.* Van Nostrand Reinhold, London.

Flemming, N.C. (ed.) (1977) *The Undersea.* Cassell, London.

Folk, R.L. (1980) *Petrology of Sedimentary Rocks.* Hemphill, Austin, Texas.

Fraser, J. (1962) *Nature Adrift: The Story of Marine Plankton.* Foulis, London.

Frazier, D.E. (1967) Recent deltaic deposits of the Mississippi River: their development and chronology. *Trans. Gulf-Coast Assn of Geol. Socs. New Orleans* **17**, 287–315.

Freeland, H.D., Farmer, D.M. and Levings, C.D. (eds.) (1980) *Fjord Oceanography.* Plenum, New York.

Friedman, G.M. and Sanders, J.E. (1978) *Principles of Sedimentology.* Wiley, Chichester.

Friedrich, H. (1969) *Marine Biology* (Transl. from German). Sidgwick and Jackson, London.

Frost, B.W. (1972) Effects of size and concentration of food particles on the feeding behaviour of the marine planktonic copepod *Calanus pacificus. Limnol. and Oceanogr.* **17**, 805–815.

Furness, R.W. and Monaghan, P. (1987) *Seabird Ecology.* Blackie, Glasgow and London.

Gage, J.D. (1985) The analysis of population dynamics in deep-sea benthos. In Gibbs, P.E. (ed.) *Proc. 19th European Marine Biologial Symp.* Cambridge University Press, Cambridge, 201–212.

Garofalo, D. (1979) Air and spacecraft remote sensing applied to coastal geomorphology. *Oceanogr. and Marine Ecol. Ann. Rev.* **17**, 43–100.

Geyer, R.A. (ed.) (1980) *Marine Environmental Pollution.* Vol. 1: *Hydrocarbons.* Elsevier, Amsterdam.

Geyer, R.A. (ed.) (1981) *Marine Environmental Pollution.* Vol 2: *Dumping and Mining.* Elsevier, Amsterdam.

Gibbs, R.J. (ed.) (1974) *Suspended Solids in Water.* Plenum Press, New York.

Giese, A.C. and Pearse, J.S. (1974) *Reproduction of Marine Invertebrates* (2 vols.) Academic Press, New York.

Goldberg, E.D. (1972) *A Guide to Marine Pollution.* Gordon and Breach, London.

Goldberg, E.D. (ed.) (1973) *North Sea Science.* MIT Press, Cambridge, Massachusetts.

Goldberg, E.D. (1974) *Marine Chemistry.* Wiley, Chichester.

Goldberg, E.D. (1976) *Strategies of Marine Pollution Monitoring.* Wiley, Chichester.

Goldberg, E.D., Steele, J.H., O'Brien, J.J. and McCave, I.M. (1977) *Marine Modelling,* Wiley, Chichester.

Gooday, A.J. (1986) Meiofaunal foraminiferans from the bathyal Porcupine Seabight (northeast Atlantic): size structure, standing stock, taxonomic composition, species diversity and vertical distribution in the sediment. *Deep Sea Research* **33**, 1345–1373.

Gordon, H.R. (1980) Phytoplankton pigments from the *Nimbus* 7 coastal zone colour scanner: comparisons with surface measurements. *Science* **210**, 63–66.

Goreau, T.F. and Hartman, W.D. (1953) Boring sponges as controlling factors in the formation and maintenance of coral reefs. In *Mechanisms of Hard Tissue Destruction,* AAAS Publ. No. 75, American Association for the Advancement of Science, Washington DC, 25–54.

Goulden, C.E. (ed.) (1977) *The Changing Scenes in Natural Sciences 1776–1976.* Academy of Natural Sciences, Philadelphia, Special Publication 12.

Gower, J.F.R. (ed.) (1981) *Oceanography from Space.* Plenum Press, New York.

Graham, M. (ed.) (1956) *Sea Fisheries: Their Investigation in the United Kingdom.* Edward Arnold, London.

Grant, A. (1985) Analysis of continuous reproduction in deep-sea seastars. In Gibbs P.E. (ed.), *Proc. 19th Marine Biol. Symp.* Cambridge University Press, Cambridge, 213–222.

Grassle, J.F. (1986) The ecology of deep-sea hydrothermal vent communities. *Advances in Marine Biology* **23**, 302–362.

Gray, J.S. (1974) Animal—sediment relationships. *Oceanogr. and Marine Biol. Ann. Rev.* **12**, 223–261.

Gray, J.S. (1981) *The Ecology of Marine Sediments.* Cambridge University Press, Cambridge.

Green, J. (1968) *The Biology of Estuarine Animals.* Sidgwick and Jackson, London.

Groen, P. (1967) *The Waters of the Sea.* Van Nostrand, London.

Guillcher, A. (1958) *Coastal and Submarine Morphology.* Methuen, London.

Gulland, J.A. (1977) *Fish Population Dynamics.* Wiley, Chichester.

Gurney, R.G. (1932) *British Fresh-water Copepoda.* (2 vols.) Royal Society, London.

Hails, J. and Carr, A. (eds.) (1975) *Nearshore Sediment Dynamics and Sedimentation.* Wiley-Interscience, London.

Hardy, A.C. (1956) *The Open Sea: The World of Plankton.* Collins, London.

Hardy, A.C. (1959) *The Open Sea:* Part II *Fish and Fisheries.* Collins, London.

Harrison, R.J. (1972) Vol. 1; (1974) Vol. 2; (1977) Vol. 3. *Functional Anatomy of Marine Mammals.* Academic Press, London.

Harrison, R.J. and Kooyman, G.L. (1971) *Diving in Marine Mammals*. Oxford Biology Readers No. 6, Oxford University Press, London.

Harvey, H.W. (1957) *The Chemistry and Fertility of Sea Waters* (2nd edn.) Cambridge University Press, London.

Harvey, J.G. (1976) *Atmosphere and Ocean*. Artemis, Sussex.

Hedgpeth, J.W. (ed.) (1957) *Treatise on Marine Geology and Paleoecology*. Geol. Soc. America Mem. 67.

Head, P.C. (ed.) (1985) *Practical Estuarine Chemistry*. Cambridge University Press, Cambridge.

Heezen, B.C. and Hollister, C.D. (1971) *The Face of the Deep*. Oxford University Press, London.

Heezen, B.C., Tharp, M. and Ewing, M. (1959) The floors of the oceans. I. The North Atlantic. *Spec. Pap. Geol. Soc. America, Washington*, 65.

Heinrich, A.K. (1962) The life histories of plankton animals and seasonal cycles of plankton communities in the oceans. *Journal du Conseil. Conseil permanent international pour l'exploration de la mer* **27**, 15–24.

Heip, C., Herman, P.M.J. and Coomans, A. (1982) The productivity of marine meiobenthos. *Academiae Analecta* **44**, 1–20.

Hela, I. and Laevastu, T. (1970) *Fisheries Hydrography*. Fishing News (Books), London.

Herman, Y. (ed.) (1974) *Marine Geology and Oceanography of the Arctic Seas*. Springer Verlag, Berlin.

Herring, P.J. and Clarke, M.R. (eds.) (1972) *Deep Oceans*. Arthur Baker, London.

Hessler, R.R. and Sanders, H.L. (1967) Faunal diversity in the deep-sea. *Deep Sea Res.* **14**, 65–78.

Hill, M.N. *et al.* (eds.) (1962–1983) *The Sea* (8 vols.) Interscience, New York.
Vol. 1 (1962) *Physical Occenography*.
Vol. 2 (1963) *Composition of Sea-water. Comparative and Descriptive Oceanography*.
Vol. 3 (1963) *The Earth beneath the Sea. History*.
Vol. 4 (1970) *New Concepts of Sea Floor Evolution*.
Vol. 5 (1974) *Marine Chemistry*.
Vol. 6 (1977) *Marine Modelling*.
Vol. 7 (1981) *The Oceanic Lithosphere*.
Vol. 8 (1983) *Deep-Sea Biology*.

Hjulström, F. (1939) Transportation of detritus by moving water. In *Recent Marine Sediments* (ed. Trask, P.D.), Murby & Co., for the American Association of Petroleum Geologists, 5–31.

Holme, N.A. and McIntyre, A.D. (1984) *Methods for the Study of Marine Benthos* (2nd edn.) Blackwell, Oxford.

Huet, M. (1971) *Textbook of Fish Culture*. Fishing News (Books), London.

Hunt, L.M. and Groves D.G. (1980) *Ocean World Encyclopedia*. McGraw-Hill, New York.

Huston, M.A. (1985) Patterns of species diversity on coral reefs. *Ann. Rev. Ecol. and Systematics* **16**, 149–177.

Hylleberg, J. (1977) *Okoligiske Problemstillinger*. Inst. fur Genetik & Okologi, Aarhus, Denmark.

Inderbitzen, A.L. (ed.) (1974) *Deep Sea Sediments: Physical and Mechanical Properties*. Plenum Press, New York.

Ingham, A.E. (ed.) (1975) *Sea Surveying* (2 vols.) Wiley, Chichester.

Ingle, J.C. (1966) *The Movement of Beach Sand. An Analysis using Fluorescent Grains*. Developments in Sedimentology, no. 5. Elsevier, Amsterdam.

Iversen, E.S. (1972) *Farming the Edge of the Sea*. Fishing News (Books), London.

Jannasch, H.W. (1985) The chemosynthetic support of life and the microbial diversity at deep-sea hydrothermal vents. *Proc. Roy. Soc. London* **B225**, 277–297.

Jannasch, H.W. and Taylor, C.D. (1984) Deep sea microbiology. *Ann. Rev. Microbiol.* **38**, 487–514.

Jefferies, R.L. and Davy, A.J. (eds.) (1979) *Ecological Processes on Coastal Environments.* Blackwell, Oxford.

Johnson, W.S., Gignon, A., Gulman, S.L. and Mooney, H.A. (1974) Comparative photosynthetic capacities of intertidal algae under exposed and submerged conditions. *Ecology* **55**, 450–453.

Johnston, R. (1977) *Marine Pollution.* Academic Press, London.

Jones, O.A. and Endean R. (eds.) (1976) *Biology and Geology of Coral Reefs* (3 vols.) Academic Press, New York.

Kennett, J.P. (1982) *Marine Geology.* Prentice Hall, Englewood Cliffs.

King, C.A.M. (1975) *Introduction to Physical and Biological Oceanography.* Edward Arnold, London.

King, C.A.M. (1975) *Introduction to Marine Geology and Geomorphology.* Edward Arnold, London.

Kinne, O. (1972) *Marine Ecology. A Comprehensive, Integrated Treatise on Life in Oceans and Coastal Waters.* Wiley-Interscience, New York.
   Vol. 1 *Environmental Factors.*
   Vol. 2 *Physiological Mechanisms.*
   Vol. 3 *Cultivation.*
   Vol. 4 *Dynamics.*
   Vol. 5 *Ocean Management.*

Koblentz-Mishke, O.J., Volkovinsky, V.V. and Kabanova, J.G. (1970) Plankton primary production of the world ocean. In *Scientific Exploration of the South Pacific*, National Academy of Sciences, Washington, DC, 183–193.

Komar, P.D. (1976) *Beach Processes and Sedimentation.* Prentic-Hall, Englewood Cliffs.

Korringa, P. (1976) *Farming Marine Organisms Low in the Food Chain.* Elsevier, Amsterdam.

Krauskopf, K.B. (1979) *Introduction to Geochemistry* (2nd edn.) McGraw-Hill, London.

Kriss, A.E. (1963) *Marine Microbiology (Deep Sea).* Oliver and Boyd, Edinburgh.

Kriss, A.E., Mishustina, I.E., Mitskevich, N., and Zemtsova, E.V. (1967) *Microbial Population of Oceans and Seas.* Edward Arnold, London.

Kuenen, P.H. (1950) *Marine Geology.* Wiley, Chichester.

Ladd, H.S. (ed.) (1957) *Treatise on Marine Ecology and Paleoecology.* vol. II: *Paleoecology. Geol. Soc. America* Mem. 67.

Laevastu, T. and Hela, I. (1970) *Fisheries Oceanography.* Fishing News (Books), London.

Lambe, W.T. Whitman, R.V. (1979) *Soil Mechanics, SI Version.* Wiley, New York.

Lang, J.C. (1973) Interspecific aggression by scleractinian corals. 2. Why the race is not only to the swift. *Bull. Mar. Sci.* **23**, 260–279.

Langmuir, I. (1938) Surface motion of water induced by wind. *Science* **87**, 119–123.

Lauff, G.H. (ed.) (1967) *Estuaries.* American Association for the Advancement of Science, Washington DC.

Lawrence, J.M. (1975) On the relationship between marine plants and sea-urchins. *Oceanogr. and Marine Biol. Ann. Rev.* **13**, 213–286.

Le Fevre, J. (1986) Aspects of the biology of frontal systems. *Adv. Marine Biol.* **23**, 164–299.

Levinton, J.S. (1982) *Marine Ecology.* Prentic-Hall, Englewood, Cliffs.

Levring, T., Hoppe, H.A. and Schmidt, O.J. (1969) *Marine Algae. A Survey of Research and Utilization.* Cram, De Gruyter, Hamburg.

Lewis, J.R. (1964) *The Ecology of Rocky Shores.* English Universities Press, London.

Lewis, J.B. (1977) Processes of organic production on coral reefs. *Biol. Revs* **52**, 305–347.

Lintz, J. and Simonett, D.S. (eds.) (1976) *Remote Sensing of Environment.* Addison-Wisley, Reading, Mass.

Livingston, D.A. (1963) Chemical composition of rivers and lakes. *Prof. Pap. US Geol. Surv.* 440 G.

Livingston, R.J. (ed.) (1979) *Ecological Processes in Coastal and Marine Systems.* Plenum Press, New York.

Lobban, C.S. and Wynne, M.J. (eds.) *The Biology of Seaweeds*. Blackwell, Oxford.

Long, S.P. and Mason, C.F. (1983) *Saltmarsh Ecology*. Blackie, Glasgow and London.

Longhurst, A.R. (ed.) (1981) *Analysis of Marine Ecosystems*. Academic Press, London.

Lubchenko, J. (1980) Algal zonation in the New England rocky intertidal community: an experimental analysis. *Ecology* **61**, 333–344.

McCave, I.N. (ed.) (1976) *The Benthic Boundary Layer*. Plenum Press, New York.

MacDonald, A.G. (ed.) (1975) *Physiological Aspects of Deep-Sea Biology*. Cambridge University Press, Cambridge.

McIntyre, A.D. (1969) Ecology of marine meiobenthos. Biol. Revs **44**, 245–290.

MacIntyre, F. (1970) Why the sea is salt. *Scientific American* **223**, 104–115.

McLellan, H.J. (1965) *Elements of Physical Oceanography*. Pergamon Press, Oxford.

McLusky, D.S. (1981) *The Estuarine Ecosystem*. Blackie, Glasgow and London.

MacNae, W. (1968) A general account of the fauna and flora of mangrove swamps and forests in the Indo-West-Pacific region. *Adv. Marine Biol.* **6**, 73–270.

McRoy, C.P. and Hellferich, C. (eds.) (1977) *Seagrass Ecosystems: a Scientific Perspective*. Marcel Dekker, New York.

Malins, D.C. (ed.) (1977) *Effects of Petroleum on Arctic and Subarctic Marine Environments and Organisms*. Academic Press, London.
Vol. 1 *Nature and Fate of Petroleum*.
Vol. 2 *Biological Effects*.

Mann, K.H. (1969) The dynamics of aquatic ecosystems. *Adv. Ecol. Res.* **6**, 1–81.

Mann, K.H. (1982) *Ecology of Coastal Waters*. Blackwell, Oxford.

Margalef, R. (1968) *Perspectives in Ecological Theory*. University of Chicago Press, Chicago.

Margalef, R. (ed.) (1985) *Key Environments. Western Mediterranean*. Pergamon Press, Oxford.

Marshall, N.B. (1979) *Developments in Deep Sea Biology*. Blandford, Poole.

Martin, D.F. (1970) *Marine Chemistry* (Vols. I, II). Marcel Dekker, New York.

Maul, G.A. (1985) *Introduction to Satellite Oceanography*. Martinus Nijhoff, Dordrecht.

Maxwell, A.E. (1971) *New concepts of Sea Floor Evolution* (Parts 1 and 2) Wiley, Chichester.

Meadows, P.S. (1964a) Experiments on substrate selection by *Corophium* species: films and bacteria on sand grains. *J. exp. Biol.* **41**, 499–511.

Meadows, P.S. (1964b) Experiments on substrate selection by *Corophium volutator* (Pallas): depth selection and population density. *J. exp. Biol.* **41**, 677–687.

Meadows, P.S. (1964c) Substrate selection by *Corophium* species: the particle size of substrates. *J. Anim. Ecol.* **33**, 387–394.

Meadows, P.S. (1986) Biological activity and seabed sediment structure. *Nature, Lond.* **323**, 207.

Meadows, P.S. and Anderson, J.G. (1966) Micro-organisms attached to marine and fresh-water sand grains. *Nature, Lond.* **212**, 1059–1060.

Meadows, P.S. and Anderson, J.G. (1968) Microorganisms attached to marine sand grains. *J. Marine Biol. Assn UK* **48**, 161–175.

Meadows, P.S. and Anderson, J.G. (1979) The microbiology of interfaces in the sea. *Progr. Industr. Microbiol.* **15**, 207–265.

Meadows, P.S. and Campbell, J.I. (1972a) Habitat selection by aquatic invertebrates. *Adv. Marine Biol.* **10**, 271–382.

Meadows, P.S. and Campbell, J.I. (1972b) Habitat selection in the marine environment: the evolution of a concept. In *2nd Int. Congr. Hist. Oceanogr. Edinb., Proc. Roy. Soc. Edinb.* **B73**, 145–157.

Meadows, P.S. and Campbell, J.I. (1978) Habitat selection and games theory. In *12th European Marine Biol. Symp.* eds. D.S. McLusky and A.J. Berry, Pergamon Press, Oxford, 289–295.

Meadows, P.S. and Ruagh, A.A. (1981) A multifactorial analysis of the behavioral responses of *Corophium volutator* (Pallas) to temperature-salinity combinations. *Marine Ecol. Progr. Ser.* **6**, 183–190.

Meadows, P.S. and Tait, J. (1985) Bioturbation, geotechnics and microbiology at the sediment-water interface in deep-sea sediments. In ed. *19th European Marine Biology Symposium*, P.E. Gibbs, Cambridge University Press, Cambridge, 191–199.

Meadows, P.S. and Tufail, A. (1986) Bioturbation, microbial activity and sediment properties in an estuarine ecosystem. *Proc. Roy. Soc. Edinb.* **B90**, 129–142.

Mero, J.L. (1964) *The Mineral Resources of the Sea*. Elsevier, Amsterdam.

Meyer, R.E. (ed.) (1972) *Waves on Beaches*. Academic Press, London.

Michanek, G. (1975) *Seaweed Resources of the Ocean*. FAO, Rome.

Milliman, J.D. (1974) *Marine Carbonates*. Springer Verlag, Berlin.

Mills, E.L. (1980) *The Structure and Dynamics of Shelf and Slope Ecosystems of the North-east Coast of North America*. Bell W. Baruch Libr. Mar. Sci., University of South Carolina Press, Carolina.

Moore, P.G. and Seed, R. (eds.) (1985) *The Ecology of Rocky Shores*. Hodder and Stoughton, London.

Morel, F.M.M. (1983) *Principles of Aquatic Chemistry*. Wiley-Interscience, New York.

Morris, S. and Taylor, A.C. (1985) The respiratory response of the intertidal prawn *Palaemon elegans* (Rathke) to hypoxia and hyperoxia. *Compar. Biochem. Physiol.* **81a**, 633–639.

Morris, S., Taylor, A.C., Bridges, C.R. and Grieshaber, M.K. (1985) Respiratory properties of the haemolymph of the intertidal prawn *Palaemon elegans* (Rathke). *J. exp. Zool.* **233**, 175–186.

Muus, K. (1966) Notes on the biology of *Protohydra leuckarti* Greef, (Hydroidea, Protohydridae). *Ophellia* **3**, 141–150.

Nairn, A.E.M. (gen. ed.) (1985) *The Ocean Basins and Margins*. Plenum Press, New York.
Vol. 1 *The South Atlantic.*
Vol. 2 *The North Atlantic.*
Vol. 3 *The Gulf of Mexico and the Caribbean.*
Vol. 4A *The Eastern Mediterranean.*
Vol. 4B *The Western Mediterranean.*
Vol. 5 *The Arctic Ocean.*
Vol. 6 *The Indian Ocean.*
Vol. 7 *The Pacific Ocean* (in two parts).

NASA (1972) *ERTS-1 Data Users' Handbook*. National Aeronautics and Space Administration, Goddard Space Flight Centre, Greenbelt, Michigan.

NCC UK (1984) *Conservation in the Marine Environment*. Nature Conservancy Council, London.

Nedwell, D.B. and Brown, C.M. (eds.) (1982) *Sediment Microbiology*. Academic Press, London.

Nelson-Smith, A. (1972) *Oil Pollution and Marine Ecology*. Elek, London.

Neumann, G. (1968) *Ocean Currents*. Elsevier, Amsterdam.

Newell, R.C. (1970) *Biology of Intertidal Animals*. Logos, London.

Newell, N.D. (1971) An outline history of tropical organic reefs. *Novitates* **2465**, 1–37.

Newell, G.E. and Newell, R.C. (1977) *Marine Plankton: a Practical Guide* (3rd edn.). Hutchinson Educational, London.

Newton, L. (1931) *A Handbook of the British Seaweeds*. British Museum (Natural History), London.

North, W.J., Neushul, M. and Clendenning, K.A. (1964) Successive biological changes observed in a marine cove exposed to a large spillage of mineral oil. *Symp. Poll. mar. Microorg. Prod. petrol.* Monaco, 335–354.

Nowell, A.R.M. and Jumars, P.A. (1984) Flow environments of aquatic benthos. *Ann. Rev. Ecol. and Systematics* **15**, 303–328.

Odum, E.P. and Smalley, A.E. (1959) Comparison of population energy flow of a herbivorous and deposit-feeding invertebrate in a salt marsh ecosystem. *Proc. Nat. Acad. Sci. USA* **45**, 617–622.

Odum, H.T., Copeland, B.J. and McMahon, E.A. (eds.) (1974) *Coastal Ecological Systems of the United States* (2 vols.) The Conservation Foundation, Washington DC.

K

256 INTRODUCTION TO MARINE SCIENCE

Odum, W.E. and Heald, E.J. (1975) The detritus-based food web of an estuarine mangrove community. In *Estuarine Research*, Vol. 1., L.E. Cronin (ed.) Academic Press, New York 265–286.

Olson, T.A. and Burgess, F.J. (1970) *Pollution and Marine Ecology*. Wiley-Interscience London.

Oppenheimer, C.H. (ed.) (1961) *Symposium on Marine Microbiology*. Thomas, Springfield, Ill

Paine, R.T. (1966) Food web complexity and species diversity. *Amer. Naturalist* **100**, 91–93

Paine, R.T. and Vadas, R.L. (1969) The effects of grazing by sea-urchins *Strongylocentrotus* spp. on benthic algal populations. *Limnol. and Oceanogr.* **14**, 710–719.

Parker, R.H. (1975) *The Study of Benthic Communities*. Elsevier, Amsterdam.

Parsons, T.R., Takahashi, M. and Hargrave, B. (1984) *Biological Oceanographic Processes* (3rd edn.) Pergamon Press, Oxford.

Patriquin, D.G. and McLung, C.R. (1978) Nitrogen accretion, and the nature and possible significance of N_2 fixation (acetylene reduction) in a Nova Scotian *Spartina alterniflora* stand. *Marine Biol.* **47**, 227–242.

Penzias, W. and Goodman, M.W. (1973) *Man Beneath the Sea*. Wiley, Chichester.

Perkins, E.J. (1974) *The Biology of Estuaries and Coastal Waters*. Academic Press, London

Perry, A.H. and Walker, J.M. (1977) *The Ocean–Atmosphere System*. Longman, London.

Phillips, O.M. (1966) *The Dynamics of the Upper Ocean*. Cambridge University Press, London

Pickard, G.L. (1975) *Descriptive Physical Oceanography* (2nd edn.) Pergamon Press, Oxford

Pielou, E.C. (1977) *Mathematical Ecology*. Wiley, New York.

Pipkin, B.W., Gorsline, D.S., Casey, R.E. and Hammond, D.E. (1977) *Laboratory Exercises in Oceanography*. Freeman, San Francisco.

Platt, T., Mann, K.H. and Ulanowicz, R.E. (eds.) (1981) *Mathematical Models in Biological Oceanography*. UNESCO, Paris.

Poutanen, E.-L. (1985) Humic and fulvic acids in marine sediments. *Finnish Marine Research* **251**, 1–45.

Press, F. and Siever, R. (1982) *Earth* (3rd ed.) Freeman, San Francisco.

Price, J.H., Irvine, D.E.G. and Farnham, W.F. (eds.) (1980) *The Shore Environment* (2 vols.) Academic Press, New York.

Ragotzkie, R.A. (ed.) (1983) *Man and the Marine Environment*. CRC Press, Boca Raton Florida.

Raymont, J.E.G. (1980) *Plankton and Productivity in the Oceans*. (2nd edn.) (2 vols.) Pergamon Press, Oxford.

Redford, A.C. (1958) The biological control of chemical factors in the environment. *Amer Scientist* **46**, 205–222.

Reineck, H.-E. and Singh, I.B. (1980) *Depositional Sedimentary Environments* (2nd edn. Springer Verlag, Berlin.

Reise, K. (1985) *Tidal Flat Ecology*. Springer Verlag, Berlin.

Rex, M.A. (1981) Community structure in the deep-sea benthos. *Ann. Rev. Ecol. and Systematics* **12**, 331–353.

Rhoads, D.C. (1967) Biogenic reworking of intertidal and subtidal sediments in Barnstable Harbor and Buzzards Bay, Massachusetts. *J. Geol.* **75**, 461–474.

Rhoads, D.C. (1974) Organism-sediment relations on the muddy sea floor. *Oceanogr. and Marine Biol. Annu. Rev.* **12**, 263–300.

Ricketts, E.F., Calvin, J. and Hedgpath, J.W. (1968) *Between Pacific Tides*. Stanford University Press, Stanford.

Riley, J.P. and Chester, R. (1971) *An Introduction to Marine Chemistry*. Academic Press London.

Riley, J.P. and Skirrow, G. (ed.) (1965) *Chemical Oceanography*. Academic Press, London.

Robinson, I.S. (1985) *Satellite Oceanography. An Introduction for Oceanographers and Remote Sensing Scientists*. Ellis Horwood, Chichester.

Round, F.E. (1981) *The Biology of the Algae*. (2nd edn.) Edward Arnold, London.

Rounsefell, G.A. (1975) *Ecology, Utilization, and Management of Marine Fisheries.* Mosby, St Louis.
Russell, F.S. and Yonge, C.M. (1963) *The Seas: Our Knowledge of Life in the Sea and How it is Gained* (2nd edn.) Frederick Warne, London.
Russell-Hunter, W.D. (1970) *Aquatic Productivity.* Macmillan, London.
Ryther, J.H. (1969) Photosynthesis and fish production in the sea. The production of organic matter and its conversion to higher forms of life vary throughout the world ocean. *Science* **166**, 72–76.
Sanders, H.L. (1968) Marine benthic diversity: a comparative study. *Amer. Naturalist* **102**, 243–282.
Sanders, H.L. and Hessler, R.R. (1969) Ecology of the deep-sea benthos. *Science* **163**, 1419–1424.
Schäfer, W. (1972) *Ecology and Paleoecology of Marine Environments.* Oliver and Boyd, Edinburgh.
Scheltema, R.S. (1971) Larval dispersal as a means of genetic exchange between geographically separated populations of Shallow-water benthic marine gastropods. *Biological Bulletin. Marine Biological Laboratory, Woods Hole, Massachusetts* **140**, 284–322.
Scheltema, R.S. (1986) On dispersal and planktonic larvae of benthic invertebrates: an eclectic overview and summary of problems. *Bull. Marine Sci.* **39**, 290–322.
Scholander, P.F. (1968) How mangroves desalinate seawater. *Physiol. Plant.* **21**, 251–261.
Scholander, P.F., van Dam, L. and Scholander, S.I. (1955) Gas exchange in the roots of mangroves. *Amer. J. Bot.* **42**, 92–98.
Scientific American Books (1969) *The Ocean.* Freeman, San Francisco.
Seibold, E. and Berger, W.H. (1982) *The Sea Floor.* Springer Verlag, Berlin.
Selley, R.C. (1982) *An Introduction to Sedimentology.* Academic Press, London.
Shannon, C.E. and Weaver, W. (1949) *The Mathematical Theory of Communication.* University of Illinois Press, Urbana.
Shepard, F.P. (1973) *Submarine Geology* (3rd edn.) Harper and Row, New York.
Shepard, F.P. (1977) *Geological Oceanography: Evolution of Coasts, Continental Margins and the Deep-Sea Floor.* Crane, Russach, New York.
Shepard, F.P., Dill, R.F. (1966) *Submarine Canyons and other Sea Valleys.* Rand McNally, Chicago.
Shirley, M.L. (ed.) (1966) *Deltas in their Geological Framework.* Houston Geological Sociery, Houston.
Shubert, L.E. (ed.) (1984) *Algae as Ecological Indicators.* Academic, London.
Simpson, E.H. (1949) Measurement of diversity. *Nature, Lond.* **163**, 688.
Simpson, J.H. (1981) The shelf-sea fronts: implications of their existence and behaviour. *Phil. Trans. Roy. Soc. London* **A 302**, 531–543.
Simpson, J.H. and Hunter, J.R. (1974) Fronts in the Irish Sea. *Nature, Lond.* **350**, 404–406.
Smayda, J.T. (1970) The suspension and sinking of phytoplankton in the sea. *Oceanogr. and Marine Biol. Ann. Rev.* **8**, 353–414.
Smith, D.L. (1977) *A Guide to Marine Coastal Plankton and Marine Invertebrate Larvae.* Kendall/Hunt, Dubuque, Iowa.
Smith, M.J. (1981) *Soil Mechanics.* (4th edn.) George Godwin, London.
Smith, J.E. (ed.) (1968) *'Torrey Canyon' Pollution and Marine Life.* Cambridge University Press, London.
Snedecor, G.W. and Cochran, W.G. (1980) *Statistical Methods* (7th edn.) Iowa State University Press, Iowa.
Snodgrass, F.E. (1968) Deep sea instrument capsule. *Science* **162**, 78–87.
Sokal, R.R. and Rohlf, F.J. (1981) *Biometry* (2nd edn.) Freeman, San Francisco.
Southward, A.J. (1965) *Life on the Sea-Shore.* Heinemann Educational Books, London.
Spicer, J.I., Moore, P.G. and Taylor, A.C. (1987) The physiological ecology of land invasion by the Talitridae (Crustacea: Amphipoda). *Proc. Roy. Soc. Lond.* In press.

Spotte, S.H. (1970) *Fish and Invertebrate Culture.* Wiley-Interscience, London.
Stanier, R.Y., Adelberg, E.A. and Ingraham, J.L. (1977) *General Microbiology* (4th edn.) Macmillan, London.
Stanley, D.J. (ed.) (1972) *The Mediterranean Sea: a Natural Sedimentation Laboratory.* Dowden, Hutchinson & Ross, Stroudsburg.
Stanley, D.J. (1976) *Marine Sediment Transport and Environmental Management.* Wiley, Chichester.
Stavn, R.H. (1971) The horizontal-vertical distribution hypothesis: Langmuir circulation and *Daphnia* distribution. *Limnol. and Oceanogr.* **16**, 453–466.
Steedman, H.F. (ed.) (1976) *Zooplankton Fixation and Preservation.* Unesco Press: Paris.
Steele, J.H. (1974) *The Structure of Marine Ecosystems.* Blackwell, Oxford.
Steele, J.H. (ed.) (1978) *Spatial Pattern in Plankton Communities.* Plenum Press, New York.
Steers, J.A. (1953) *The Sea Coast.* Collins, London.
Steers, J.A. (ed.) (1971) *Applied Coastal Geomorphology.* Macmillan, London.
Steers, J.A. (ed.) (1971) *Introduction to Coastline Development.* Macmillan, London.
Stehli, F.G., McAlester, A.L. and Helsley, C.E. (1967) Taxonomic diversity of recent bivalves and some implications for geology. *Geol. Soc. Amer. Bull.* **78**, 455–466.
Stephenson, T.A. and Stephenson, A. (1972) *Life between Tidemarks on Rocky Shores.* Freeman, San Francisco.
Stern, M.E. (1975) *Ocean Circulation Physics.* Academic Press, London.
Stevenson, L.H. and Colwell, R.R. (eds.) (1973) *Estuarine Microbial Ecology.* University of South Carolina Press, Columbia.
Stewart, R.H. (1985) *Methods of Satellite Oceanography.* University of California Press, Berkeley.
Stoddart, D.R. and Yonge, C.M. (eds.) (1971) *Regional Variation in Indian Ocean Coral Reefs* (35th Symp. Zool. Soc. London.) Academic Press, London.
Strahler, A.N. and Strahler, A.H. (1983) *Modern Physical Geography* (2nd edn.) Wiley, New York.
Strathmann, R.R. (1985) Feeding and nonfeeding larval development and life-history evolution in marine invertebrates. *Ann. Rev. Ecol. Syst.* **16**, 339–361.
Sundborg, A. (1956) The River Klarälven. A study of fluvial processes. *Geografiska Annaler* **38**, 127–316.
Sverdrup, H.U., Johnson, M.W. and Flemming, R.H. (1942) *The Oceans. Their Physics, Chemistry and General Biology.* Prentice-Hall, New York.
Swedmark, B. (1964) The interstitial fauna of marine sand. *Biol. Rev.* **39**, 1–42.
Szekielda, K.-H. (1976) Spacecraft Oceanography. *Oceanogr. and Marine Biol. Ann. Rev.* **14** 99–166.
Tait, R.V. (1981) *Elements of Marine Ecology* (3rd edn.) Butterworths, London.
Taylor, A.C. (1987) The ecophysiology of Decapods in the rock pool environment. In *Aspects of Decapod Crustacean Biology,* (eds.) Fincham, A.A. and Rainbow, P.S.) *Zool. Soc. London Symp.* In press.
Taylor, J.H. (1964) Some aspects of diagenesis. *The Advancement of Science* **20**, 417–436.
Teal, J.M. (1962) Energy flow in the salt marsh ecosystem of Georgia. *Ecology* **43**, 614–624.
Tenore, K.R. and Coull, B.C. (eds.) (1980) *Marine Benthic Dynamics.* University of South Carolina Press, Columbia.
Tevesz, M.J.S. and McCall, P. (eds.) (1983) *Biotic Interactions in Recent and Fossil Benthic Communities.* Plenum Press, New York.
Thorson, G. (1946) Reproduction and Larval Development of Danish Marine Bottom Invertebrates, with Special Reference to the Planktonic Larvae in the Sound (Oresund) *Meddelelser fra Kommissionen for Danmarks Fiskeri-og Havundersøgelser,* Serie: Plankton Bd. 4.
Thorson, G. (1971) *Life in the Sea.* Weidenfeld and Nicolson, London.
Thurman, H.V. and Webber H.H. (1984) *Marine Biology.* Merrill, Columbus, Ohio.

Tilman, D. Kilham, S.S. and Kilham, P. (1982) Phytoplankton community ecology: the role of limiting nutrients. *Ann. Rev. Ecol. and Systematics* **13**, 349–372.

Tressler, D.K. (1940) *Marine Products of Commerce*. Reinhold, New York.

Trueman, E.R. (1975) *The Locomotion of Soft-bodied Animals*. Edward Arnold, Bristol.

Tufail, A. (1985) Microbial aggregates on sand grains in enrichment media. *Botanica Marina* **28**, 209–211.

Tufail, A. (1987) Microbial communities colonising nutrient-enriched marine sediment. *Hydrobiologia* **148**, 245–255.

Turekian, K.K. (1968) *Oceans*. Prentice-Hall, New Jersey.

Tyler, P.A., Grant, A., Pain, S.L. and Gage, J.D. (1982) Is annual reproduction in deep sea echinoderms a response to variability in their environment? *Nature, Lond.* **300**, 747–750.

Tyler, P.A., Muirhead, A. and Colman, J. (1985) Observations on continuous reproduction in large deep-sea epibenthos. In Gibbs, P.E. (ed.). *Proc. 19th European Marine Biology Symp.*, Cambridge University Press, Cambridge, 223–230.

UNESCO/SCOR (1978) *Biogeochemistry of Estuarine Sediments*, Proc. UNESCO/SCOR workshop, 1976. UNESCO, Paris.

Valentine, J.W. (1973) *Evolutionary Ecology of the Marine Biosphere*. Prentice-Hall, Englewood Cliffs.

Valiela, Ivan (1984) *Marine Ecological Processes*. Springer Verlag, Berlin.

Vermeij, G.J. (1971) Substratum relationships of some tropical Pacific intertidal gastropods. *Marine Biol.* **10**, 315–320.

Vernberg, F.J. and Vernberg, W.B. (eds.) (1974) *Pollution and Physiology of Marine Organisms*. Academic Press, London.

Vogel, S. (1981) *Life in Moving Fluids. The Physical Biology of Flow*. Willard Grant, Boston, Massachusetts.

von Brandt, A. (1972) *Fish Catching Methods of the World*. Fishing News (Books), Surrey.

Vollenweider, R.A. (ed.) (1974) *A Manual on Methods for Measuring Primary Production in Aquatic Environments* (2nd edn.) IBP Handbook no. 12, Blackwell, Oxford.

Walton Smith, F.G. (ed.) (1974) *Handbook of Marine Science* (2 vols.) CRC Press, Cleveland.
Section I. *Oceanography*
Vol. 1 (1974) *Physical-chemistry, physics, geology, engineering.* (ed. F.G. Walton Smith)
Vol. 2 (1974) *Biological-primary productivity, plankton, general yields.* (eds.) F.A. Kalber and F.G. Walton Smith)
Section II. *Marine Products*
Vol. 1 (1976) *Compounds from Marine Organisms* (Eds. J.T. Baker and V. Murphy)
Section III. *Mariculture*
Vol. 1 (1976) and 2 (1977) (ed. G.A. Rounsefell)
Section IV. *Fisheries*
Vol. 1 (1977) (ed. G.A. Rounsefell)

Weatherly, A.H. (1972) *Growth and Ecology of Fish Populations*. Academic Press, London.

Webb, J.E. (1975) *Guide to the Marine Stations of the North Atlantic and European Waters*. The Royal Society, London.
Part 1 *Northern Europe and the East Atlantic Coast.*
Part 2 *Mediterranean Region.*
Part 3 *West Atlantic Coast and Caribbean.*

Weber, R.E. and Hagerman, L. (1981) Oxygen and carbon dioxide transporting qualities of hemocyanin in the hemolymph of a natant decapod *Palaemon adspersus. J. Compar. Physiol.* **145**, 21–27.

Werner, D. (ed.) (1977) *The Biology of Diatoms*. Blackwell, Oxford.

West, R.G. (1968) *Pleistocene Geology and Biology*. Longman, London.

Wethey, D.S. (1986) Ranking of settlement cues by barnacle larvae: influence of surface contour. *Bull. Mar. Sci.* **39**, 393–400.

Wetzl, R.G. (1975) *Limnology*. Saunders, Philadelphia.

Weyl, P.K. (1970) *Oceanography. An Introduction to the Marine Environment.* Wiley, Chichester.

Whitney, D.E., Woodwell, G.M. and Howarth, R.W. (1975) Nitrogen fixation in Flax Pond: A Long Island salt-marsh. *Limnol. and Oceanogr.* **20**, 640–643.

Whitaker, J.H.McD. (ed.) (1976) *Submarine Canyons and Deep-Sea Fans.* Dowden, Hutchinson and Ross, Stroudsburg, Pennsylvania.

Whittard, W.F. and Bradshaw, R. (ed.) (1965) *Submarine Geology and Geophysics.* Butterworths, London.

Whitten, D.G.A. and Brooks, J.V.R. (1978) *The Penguin Dictionary of Geology.* Allen Lane, London.

Wickstead, J.H. (1976) *Marine Zooplankton.* Edward Arnold, London.

Wiegel, R.L. (1964) *Oceanographical Engineering.* Prentice-Hall, New Jersey.

Wilbur, C.G. (1983) *Turbidity in the Aquatic Environment.* C.C. Thomas: Springfield, Illinois.

Wiley, M. (ed.) (1977) *Estuarine Processes* (2 vols.) Academic Press, New York.

Williams, P.J. LeB. (1981) Incorporation of microheterotrophic processes into the classical paradigm of the planktonic food web. *Kieler Meeresforsch.* **5**, 1–28.

Wimpenny, R.S. (1966) *The Plankton of the Sea.* Faber and Faber, London.

Winberg, G.G. (ed.) (1971) *Methods for the Estimation of Production of Aquatic Animals* (transl. from Russian). Academic Press, London.

Winn, H.E. and Olla, B.L. (1972) *Behaviour of Marine Animals* (2 vols.) Plenum Press, New York.

Wood, E.J.F. (1965) *Marine Microbial Ecology.* Chapman and Hall, London.

Wood, E.J.F. (1967) *Microbiology of Oceans and Estuaries.* Elsevier, Amsterdam.

Wood, E.J.F. and Johannes, R.E. (eds.) (1975) *Tropical Marine Pollution.* Elsevier, Amsterdam.

Woods, J.D. and Lythgoe, J.N. (1971) *Underwater Science. An Introduction to Experiments by Divers.* Oxford University Press, London.

Woodin, S.A. (1974) Polychaete abundance patterns in a marine soft-sediment environment: the importance of biological interactions. *Ecol. Monogr.* **44**, 171–187.

Woodin, S.A. (1978) Refuges, disturbance and community structure: a marine soft-bottom example. *Ecology* **59**, 274–284.

Woodwell, G.M., Whittaker, R.H., Reiners, W.A., Likens, G.E., Delwiche, C.C. and Botkin, D.B. (1978) The biota and the world carbon budget. *Science* **119**, 141–146.

Yen, T.F. (1977) *Chemistry of Marine Sediments.* Wiley, Chichester.

Yonge, C.M. (1949) *The Sea Shore.* Collins, London.

Zeitzschel, B. (ed.) (1973) *The Biology of the Indian Ocean.* Chapman and Hall, London.

Zenkevitch, L. (1963) *Biology of the Seas of the USSR* (transl. by S. Botcharskaya). Allen and Unwin, London.

Zenkovich, V.P. (1958) *Shores of Black and Azor Seas.* State Geographic Editions, Moscow.

Zenkovich, V.P. (1967) *Processes of Coastal Development.* Oliver and Boyd, Edinburgh.

ZoBell, C.E. (1946) Studies on redox potential of marine sediments. *Bull. Amer. Assn. Petrol. Geol.* **30**, 477–513.

ZoBell, C.E. (1946) *Marine Microbiology. A Monograph of Hydrobacteriology.* Chronica Botanica, Waltham, Massachusetts.

Marine review journals

For informative reviews and specialist areas of marine research.

Advances in Hydroscience

Advances in Marine Biology. Academic Press.

Oceanography and Marine Biology, Annual Review. Allen and Unwin.

Progress in Oceanography. Pergamon Press.

European Symposia on Marine Biology

(arranged in chronological order)

Kinne, O. and Aurich, H. (ed.) (1967) 1st Symposium: 'Experimental ecology, its significance as a marine biological tool. Subtidal ecology particularly as studied by diving techniques.' *Helgoländer Wissenschaftliche Meeresuntersuchungen*, **15**: 721 pp.

Brattström, H. (ed.) (1968) 2nd Symposium: 'The importance of water movements for biology and distribution of marine organisms.' *Sarsia*, **34**: 398 pp.

Soyer, J. (ed.) (1971) 3rd Symposium: 'Biologie des sédiments meubles. Biologie des eaux à salinité variable.' *Vie et Milieu*: Suppl. *22*, Vol. I, 1–464, Vol. 2: 464–857.

Crisp, D.J. (ed.) (1971) 4th Symposium: *Larval biology. Light in the marine environment.* Cambridge University Press: London.

Battaglia, B. (ed.) (1972) 5th Symposium: *Evolutionary aspects of marine biology. Factors affecting biological equilibria in the Adriatic brackish water lagoons.* Piccin Editore: Padova.

Zavodnik, D. (ed.) (1971) 6th Symposium: 'Productivity in coastal area of the sea. Dynamics in benthic communities.' *Thalassia* Jugslavia, **7**(i): 445 pp.

De Blok, J.W. (ed.) (1973) 7th Symposium: 'Mechanisms of migration in the marine environment. Respiratory gases and the marine organism.' *Netherlands Journal of Sea Research*, **7**: 505 pp.

Bonaduce, G.and Carrada, G.C. (ed.) (1975) 8th Symposium: 'Reproduction and sexuality in the marine environment.' *Publicazioni della Stazione Zoologica di Napoli*, **39** Suppl. 1: 727 pp.

Barnes, H. (ed.) (1975), 9th Symposium: *The biochemistry, physiology, and behaviour of marine organisms in relation to their ecology.* Aberdeen University Press: Aberdeen.

Persoone, G. and Jaspers, E. (ed.) (1976) 10th Symposium: *Population dynamics of marine organisms in relation with nutrient cycling in shallow waters.* Universa Press: Wettern, Holland.

Keegan, B.F., O'Ceidigh, P. and Boaden, P.J.S. (ed.) (1977) 11th Symposium: *Biology of benthic organisms*, Pergamon Press, Oxford.

McLusky, D.S. and Berry, A.J. (ed.) (1978) 12th Symposium: *Physiology and behaviour of Marine organisms.* Pergamon Press: Oxford.

Naylor, E. and Hartnoll, R.G. (ed.) (1979) 13th Symposium: *Cyclic phenomena in marine plants and animals.* Pergamon Press, Oxford.

Kinne, P. and Bulnheim, H.-P. (ed.) (1980) 14th Symposium: Protection of life in the sea. *Helgoland. wiss. Meeresuntersuch.* **33**: 732 pp.

Rheinheimer, G., Flugel, H., Lenz, J. and Zeitzschel, B. (ed.) (1981) 15th Symposium: lower organisms and their role in the food web. *Kieler Meeresforschungen*, Sonderheft, Nr. **5**: 588 pp.

De Blok, J.W. (ed.) (1982) 16th Symposium: Dynamic aspects of marine ecosystems. *Neth. J. Sea Res.* **16**: 505 pp.

Cabioch, L. (ed.) (1983) 17th Symposium: Fluctuation and succession in marine ecosystems. *Oceanologica Acta*, Vol. Special: 225 pp.

Gray, J.S. and Christiansen, M.E. (ed.) (1985) 18th Symposium: *Marine biology of polar regions. Effects of stress on marine organisms.* Wiley, Chichester.

Gibbs, P.E. (ed.) (1985) 19th Symposium: *Production at boundary systems. Dynamics of deep sea life. Community organisation in the benthos. Adaptive aspects of physiological and biochemical variability.* Cambridge University Press, London.

Marine science abstracting journals
These contain titles and sometimes abstracts of recently published research papers in scientific journals:

Aquatic Biology Abstracts	1
Biological Abstracts	2
Current Contents	2, 3
Marine Science Contents F. A. O.	1, 3
Science Citation Index	2, 3
Zoological Record	2, 3

1, Marine and freshwater, 2, general, including marine; 3 titles only.

Computer search facilities
There are also specialised computer facilities for searching the scientific literature. The British BLAISE and international MEDLARS and MEDLINE are available in many of the larger libraries in academic institutions.

Scientific journals publishing original research in marine science
We have divided the large number of journals into those of major interest and others. This division is somewhat arbitrary, but it is hoped that it will help the reader.
(a) Major Journals
  Annales Biologiques. Counseil International l'Exploration de la Mer, Copenhague
  Australian Journal of Marine and Freshwater Research
  Biological Bulletin, Marine Biological Laboratory, Woods Hole, Mass.
  Bulletin de l'Institut Océanographique, Monaco
  Bulletin of Marine Ecology, Edinburgh
  Bulletin of Marine Science, Florida
  Canadian Journal of Fisheries and Aquatic Science
  Cahiers de Biologie Marine
  Deep-Sea Research and Oceanographical Abstracts
  Estuarine and Coastal Shelf Science
  Fishery Investigation, M.A.F.F., London
  Helgolander wissenschaftliche Meeresuntersuchugen Hydrobiologia
  International Revue Gasainten Hydrobiologie
  Journal du Conseil
  Journal of Experimental Marine and Ecology
  Journal of Geophysical Research
  Journal of Fisheries Research Board of Canada
  Journal of the Marine Biological Association of the United Kingdom
  Journal of Marine Research
  Limnology and Oceanography
  Marine Biology
  Marine Chemistry
  Marine Ecology Progress Series
  Marine Geology
  Marine Pollution Bulletin
  Marine Research, D.A.F.F., Scotland
  Nature
  Netherlands Journal of Sea Research
  New Zealand Journal of Marine and Freshwater Research, Sarsia, Bergen
  Offshore Technology, Scientific Surveys (Offshore) Ltd., London
  Ophelia
  Pacific Science
  Science
  Tethys
  Veroffentlichungen des Instituts für Meereforschung in Bremerhaven
  Vie et Milieu, Bulletin du Laboratoire Arago:
    Série A—*Biologie Marine*
    Série B—*Océanographie*, Banyuls-sur-Mer
(b) Others
  Anales del Instituto de Biologia, Universidad Autonoma de Mexico, Serie Ciencias del Mary
    Limnologia, Mexico
  Annales de l'Institut Océanographique, Monaco

Archiv für Fischereiwissenschaft, Berlin
Archivio di Oceanografia e Limnologia, Venezia
Beiträge zur Meereskunde, Berlin
Berichte der Deutschen wissenschaftlichen Kommission für Meeresforschung, Hamburg
Bollettino di Pesca, Piscicoltura e Idrobiologia, Roma
Botanica Marina, International Review for Seaweed Research and Utilization, Hamburg
Bulletin Far Sea Fisheries Research Laboratory, Shimizu-shi
Bulletin of the Hokkaido Regional Fisheries Research Laboratory, Hokkaido
Bulletin de l'Institut Océanographique, Monaco
Bulletin of the Japan Sea Regional Fisheries Research Laboratory, Niigata
Bulletin of the Japanese Society of Scientific Fisheries, Tokyo
Bulletin de la Société Phycologique de France, Paris
Cahiers ORSTOM—Hydrobiologie, Bondy
Cahiers ORSTOM—Hydrobiologie, Bondy
Cahiers ORSTOM—Océanographie, Bondy
Contributions in Marine Science, University of Texas
Deutsche Hydrographische Zeitschrift, Hamburg
Fischerei-Forschung, Rostock
Fishery Bulletin, National Oceanic and Atmosphere Administration, National Marine Fisheries Service, Seattle, Washington
Gulf Research Reports
Indian Journal of Fisheries
Indian Journal of Marine Sciences
International Hydrographic Review, Monaco
Investigacion Pesquera, Barcelona
Journal of the Marine Biological Association of India
Journal of the Oceanographical Society of Japan
Journal of Physical Oceanography
Journal of the Tokyo University of Fisheries
Kieler Meeresforschungen, Kiel
Marine Geophysical Researches, Dordrecht, Netherlands
Marine Geotechnology
Marine Technology Society Journal
Meddelelser fra Danmarks Fiskeri-og Havundersogelser, Ny Series
Merentutkimuslaitoksen Julkaisu Havsforskningsinstitutets Skrift, Finland
Ocean Engineering
Ocean Management
Oceanography and Meterology, Nagasaki
Oceanologia, Poland
Offshore Technology, London
Okeanologiia. SSSR
Phycologia. International Phycological Society, Odense
Physis, Buenos Aires
Pubblicazioni della Stazione Zoologica di Napoli
Rapports et Procès-verbaux des réunions. Conseil International pour l'Exploration Scientifique de la Mer Méditerranée. Copenhague
Records of Oceanographic Works in Japan, Tokyo
Revista de Biologia Marina, Valparaiso
Revue Algologique, Centre National de la Recherche Scientifique, Paris
Revue Internationale d'Océanographie Médicale, CERBOM, Nice
Revue des Travaux de l'Institut des Pêches Maritime, Nantes
Sea Technology
Travaux du Centre de Recherches et d'Etudes Océanographiques, Paris

Trudy Instituta Okeanologii, USSR

Trudy. Poliarnyi nauchno-issledovatel's skii i Proektnyi Institut Morskogo, Rybnogo

Trudy. Poliarnyi nauchno-issledovatel'skii i Proektnyi Institut Morskogo Rybnogo Khoziaistva i Okeanografii imeni N.M. Knipovicha (PINRO)

Trudy. Vsesoiuznyi nauchno-issledovatel'skii Institut Morskogo Rybnogo Khoziaistva i Okeanografii (VNIRO)

Index